CW01004079

Lectures on
Structure and Significance of Science

H. Mohr

Lectures on

Structure and Significance of Science

Springer-Verlag
New York Heidelberg Berlin

H. Mohr

Biologisches Institut II der Universität
Schänzlestrasse 1
7800 Freiburg
Federal Republic of Germany

With 22 figures

Library of Congress Cataloging in Publication Data
Mohr, Hans, 1930–
Lectures on structure and significance of science.

Bibliography: p.
Includes index.
1. Science—Addresses, essays, lectures. 2. Science—Philosophy—Addresses, essays, lectures. I. Title.
Q171.M687 501 76-51383

Printed in the United States of America
9 8 7 6 5 4 3 2 1

ISBN 0-387-08091-0 Springer-Verlag New York
ISBN 3-540-08091-0 Springer Verlag Berlin Heidelberg

Acknowledgements

The opportunity of writing this treatise was made possible by a Visiting Professorship granted to me by the University of Massachusetts during the fall term of 1975. The present text is based on a series of 15 lectures that I delivered at U Mass. I am grateful to the students, and to my colleagues and friends in the Departments of Botany and Philosophy at U Mass, for their cordial reception, continuous interest, and constructive criticism. It was the positive response of my class and the fascinating intellectual climate at Amherst that encouraged me to prepare the lectures for print.

Two of my colleagues in Freiburg, the evolutionary biologist, Günther Osche, and the biophysicist, Eberhard Schäfer, have carefully read the first draft of the manuscript. Their criticisms and suggestions have been considered in the final version. My American secretary, Barbara Hoffmann, has checked the manuscript with respect to grammar and style.

This book is dedicated to Erwin Bünning and to Walter Kossel. In 1953, Bünning's book *Theoretische Grundfragen der Physiologie* was an important determinant in my decision to become a biologist. My Ph.D. work with Erwin Bünning has been the decisive experience in my scientific life. The late Walter Kossel introduced me to physics. He was not only a great physicist but also a fascinating philosopher and an admirable personality.

Freiburg i. Breisgau, March 1977 H. Mohr

Contents

To the Reader

I am not a professional philosopher of science. Rather, I am a natural scientist with some interest in the nature of scientific thought and in the significance of science. In the following lectures, I will make use of my (limited) knowledge about the sociology of science and of my (equally limited) knowledge about philosophy and history of science. The main source, however, on which I depend is my own personal experience as a practicing scientist. I hope that the professional philosophers will forgive me if my treatise does not always respect the conventional division of labor between science and philosophy. I fully agree with David Hull who recently criticized some noted scientists who tried their hands at "philosophizing": "Just as scientists are entitled to established standards of competence for their undertakings, philosophers have a right to expect at least minimal competence in theirs (1)". On the other hand, however, I feel that it is legitimate to base a reflection about the "Structure and Significance of Science" primarily on the self-understanding of the practicing scientist. My deep-rooted respect for philosophy in toto and for epistemology in particular will hopefully prevent me from becoming chauvinistic in favor of the scientific world view.

In the prologue *(Science and Responsibility)*, the problems will be formulated in broad terms. In the following lectures, the chief issues will be treated in more detail. While there is a well-established tradition of scholarship in the treatment of the "Scientific Method," [For an excellent recent collection of texts on the philosophy of science see: KRÜGER (2)] the *responsibility* of the scientist has only recently become a starting point in considering the phenomenon of science; an interesting recent experiment has been published by Michaelis and Harvey (3). The influential Anglo-Saxon schools in the philosophy of science have generally equated philosophy with epistemology, treating ethics as not properly part of academic philosophy. Since I am not obliged to any epistemological school but look at the problems from the point of view of a practicing scientist, I will not follow the tendency of excluding anything from consideration that might raise moral problems. Rather, I intend to emphasize this aspect. I will further take the liberty to look at some traditionally epistemological problems, such as empirism and rationalism, from the point of view of scientific knowledge.

Another point is that most philosophers of science, in particular within the dominant positivist school, take the Comtean view of physics as the paradigmatic science and of biology as a relatively immature and secondary study. Even as a biologist who is proud of biology, I cannot ignore this tendency since there is some truth in it.

While there is no principal difference between physics and biology, the general approach in both fields and the nature of physical and biological theories and laws obviously differ to a considerable extent. I will often refer to physics rather than to biology not only for the sake of simplicity, clarity, and brevity, but also for the reason that physics has a far wider scope than biology. Physics deals with the properties of all matter whereas biology is only concerned with living systems or with ecosystems in which living systems play the major part. All living systems are physical objects, but only a very small number of physical objects are considered to be living systems.

In the following I will often use the term "responsibility." This term implies, and I want to emphasize this at the very beginning, that we *are* responsible for our acts. I indeed presume that moral responsibility is part of human nature, irrespective of the century-long discussion on determination, free will, and moral responsibility. *Determination* to a scientist conveys the general proposition that every event has a cause. Whether this general proposition is true is a difficult question to decide, but it is certainly assumed to be true by most scientists. Otherwise science, in particular prediction, explanation, and purposive action, would not be possible. On the other hand, however, we presume that we are responsible for our acts. It is implied as a matter of course that moral responsibility is an integral part of human nature. Indeed, we all believe that moral responsibility is real.

Since moral responsibility implies free will and self-determination (in the sense that we can create de novo determinants for our conduct and thus break causal continuity), the very serious and difficult question arises whether moral responsibility (which implies free will) is compatible with our scientific knowledge, which plainly says that the concept of a breach in causal continuity is not acceptable. From the point of view of science the reality of free will cannot be conceded. On the other hand, as human beings, we depend on the belief that at least some of our actions (called "willed actions") are preceded by *deliberation* and *choice* and that our choice can be influenced by consideration of consequences. I will leave this question open at the moment. We will return to this problem when we try to analyze the nature of the principle of causality (6th Lecture) and the structure of teleological action (7th Lecture).

I suppose in the following treatise that the discoveries of science had a profound effect on man's philosophy, ethics, and spiritual beliefs. However, science still has a long way to go to reach its major goal. In the words of Sir George Porter (4)

> The highest wisdom has but one science, the science of the whole, the science explaining the Creation and man's place in it.

1st Lecture

Prologue
Science and Responsibility

The last time I talked to this audience (1), I gave a series of scientific lectures, that is, I presented a hypothesis and cases that supported the hypothesis, and I pointed out that there are no data known at present that do not agree with the hypothesis. In other words, I tried to follow carefully the established manner of dealing with a scientific problem. This time I shall not discuss a scientific problem, but I wish to consider certain aspects of science itself. When a scientist tries to contribute to the discussion on this level, he finds himself in a difficult position, for it is hardly possible to deal with the problem of science in precisely the same way that we deal with a scientific problem. To say it frankly and in advance: any philosophy of science is less rational than science itself.

One of the main problems in this connection has been that we traditionally use *verbal* language—rather than a formal language—when we talk about science and the links of science to society. Every verbal language contains a great number of semantic overtones. While such semantic overtones are the very essence of poetry and literary prose and the stock-in-trade of the political orator, verbal language is an awkward tool for the scientist who is trying to express himself without also evoking an emotional response. The words have all been used before, often in emotional situations, and the word recalls the emotion whether one wants it to or not (2). Science is a late product of cultural evolution. It emerged slowly from a welter of natural mental activity. Today science can be defined operationally as a disciplined and systematic attempt of the human mind that aims at genuine knowledge. Genuine scientific knowledge is expressed as true propositions: singular propositions (or data or facts) and general propositions (e.g., laws). True general propositions are those that describe the behavior of scientific models that are satisfying, insofar as they refer to the behavior of real systems: living or nonliving systems—those existing in nature and those made by man. The criterion of truth with respect to scientific propositions is the prediction of which is correct when compared with subsequent experience.

My thesis is that the major responsibility of the scientist, the responsibility of the scientific community, is to guarantee the truth (and therewith the reliability) of scientific propositions. This, of course, is the *traditional* responsibility of the scientific community.

Every experienced member of the scientific community knows exactly the difference between true and false propositions, and every experienced scientist knows how easy it is to make statements and how difficult it is to prove that the propositions are true. However, I would like at least to give precision to the term "law" in this context: the highest degree of scientific truth is reached when laws can be formulated, universal or limited laws—universal, if they apply to every real system, and limited, if their validity is restricted to certain classes of systems. The first and second laws of thermodynamics are famous examples of universal laws; on the other hand, the proposition $\Delta G \neq 0$ is a thermodynamic law that is applicable only to open systems, including living systems. For living systems, $\Delta G \neq 0$ is a universal law, and so on.

The next aspect I would like to consider concerns the significance of authority in science. The attitude of questioning the established authority has been as essential for the ascent of science as improved intellectual and material methods. In principle, the scientist recognizes no authority except an empirical observation of nature. In practice he will often accept a report of such an observation by a single established or distinguished researcher, but in general a consensus of several researchers is preferred. In any case, for the scientist empirical observation is the ultimate court of appeal that must be invoked if there is even the slightest doubt with regard to any statement regarding his science. The true propositions of science do not depend on any *political* authority or ideology. Scientific thinking can be suppressed by crude political power, as, for example, during the Lyssenko era in Russia (3,4), but as long as science is permitted and supported in a society, it is independent of the particular political system or ideology. The structure and the modus procedendi of science is amazingly independent of the prevailing political ideology (5,6).

Now let me return to my previous thesis, that the goal of science is genuine knowledge about real systems. The responsibility of a scientist in relation to science follows from this definition of the goal of science: the moral obligation of a scientist is to contribute to the wealth of genuine knowledge under all circumstances, even under economic or political pressure, and by doing so to decrease the amount of ignorance, prejudice, and superstition about nature, including man.

It is part of the established ethics of science that objective knowledge is good, i.e., objective knowledge has a higher value than ignorance and prejudice under all circumstances. Any "Index of Forbidden Knowledge" cannot be reconciled with the ethical foundations of science.

Two problems will arise immediately. The first problem is that every scientist is a member of the world-wide scientific community and at the same time a citizen of a country, that is, an element in a legal and political system. The latent problem becomes crucial when moral loyalty to the scientific community and loyalty to the political system collide. This conflict of loyalties arises automatically if absolute intellectual freedom, an indispensable prerequisite of scientific thinking, is no longer guaranteed in a political system. This is the case in every country where a *single* political ideology, a *specific* political dogma, rules the scene, as for instance in those countries ruled at present by dogmatic marxism. However (and this brings me to the next problem), in the modern world technology is almost exclusively, although indirectly, based on singular or general scientific propositions, or, in simpler terms, on the intellectual framework derived from genuine scientific knowledge. No country can afford to base technology on political dogma, ideology, agnostic superstition, or prejudice, even for a limited time. Modern technology can only be based on objective genuine knowledge—on the paradigms of science. This, of course, is the main reason why at least some scientific research is supported by every country whose existence requires sophisticated technology. And this is the reason why no political power can afford to suppress the freedom of thinking in the pursuit of a particular scientific research program, even if most of the country's activities are controlled by rigid ideology and the research topics themselves are selected by some central committee. Each time I visited Russia or East Germany I felt that our fellow scientists there live in a sort of natural park provided with an impenetrable fence. Inside the fence there is full intellectual freedom, but outside the fence there is none.

The relationship between science and *technology* has recently been confused by a number of misconceptions. I hope to avoid these difficulties by defining science and technology operationally. Let me first repeat my operational definition of science: the goal of science is to gain genuine knowledge about the real world, in brief, to understand and comprehend the real world. On the other hand, the goal of every technology is to change or improve the real world for the sake of man, or more precisely, for the sake of some men. Both aspects are always present in a particular research project in science or technology, even though the researcher is concerned with only one aspect. Every true scientific proposition can potentially be used technologically; and every technological research project may contribute to the progress of science. Einstein was directly concerned with basic scientific research when he postulated and tested the model leading to the equation $E = mc^2$; yet this same equation is useful in the production of nuclear bombs as well as in the production of power from nuclear sources. The Manhattan Project was directly concerned with the development of a weapon, but also contributed greatly to the development of science.

We all feel that technology can be misused. This has always been the case since man learned to master, more or less, the forces of nature. The Greeks and Romans were able, using a primitive and purely empirical technology, to damage the Mediterranean landscape to such an extent that eventually a great deal of the original soil was eroded. The rapid decay of the Golden Age of the Greeks and the more gradual decline of the Roman Empire were at least in part caused by the irreversible destruction of the natural resources of wood, soil, and water (7). The destructive potential of modern technology is tremendous compared to the primitive empirical technology of the classical and medieval ages, since this technology is based on genuine knowledge about the real world. The question arises of who is responsible for the use and misuse of technology—scientists? or if not scientists, then who else? Before I go into this matter in more detail I would like to make three firm statements:

Firstly, we have become accustomed to discussing the regressive phenomena of technology so exclusively that we have nearly forgotten how much we owe to technology. Many fancy statements about the hazards of technology show that the authors do not have the slightest idea of the circumstances under which our ancestors lived, how they suffered and how they died. The participants in the wild discussion about the hazards of technology must learn to observe two elements that are essential for every fruitful discussion: fairness and competence.

Secondly, the favorite daydream of the bewildered citizen of stopping science, of calling a moratorium on science, is quite unrealistic; but even supposing it could be accomplished, such a solution of the problem would necessarily and rapidly lead to an irreversible decay of human culture. This statement is justified because the inevitable regressive phenomena of our technological culture can only be overcome by new technology. There is no way back. We have changed the world too much already. We have about 3.5 billion people too many on this planet to follow the romantic dream (even if it were attractive) of returning to the simple life close to nature. To repeat: as the situation is, there is no way back. The regressive phenomena of present-day technology, including pollution, can only be solved by progressive technology, and not by anything else. To call a moratorium on science would be to commit suicide, since the genuine knowledge that we need for the refined and improved technology of tomorrow, especially for a humanistic biological technology, is not yet available. It can only be elaborated by the efforts of science. On this basis alone any idea of a literal standstill in science is wrong-headed and cannot be taken seriously. Moreover, no scientist, and I hope no one who cares for the growth of knowledge, would accept a moratorium, even if it were practicable throughout a politically split world. The tradition of the free mind, that is, free inquiry and free publication, has been essential in setting the standards of truth in science; it has already been eroded by secrecy in government and industry, and we need to resist any extension of this (8). Any kind of "Index of Forbidden Knowledge" would lead us back to the

Middle Ages, would cast doubt on the maturity of the human mind, would subject the finest product of cultural evolution, genuine knowledge, to the control of individual prejudice, *superstition*, or collective prejudice, *ideology*. It would be a path of no return. The tradition of free inquiry and free publication may not be interrupted.

Thirdly, despite intense scrutiny, the processes and the "laws" (if any) governing the contributions of science to economic growth and improvement have thus far remained virtually uncharted. In any case, there are no simple functional relationships between the two areas. As an example, it became obvious that contrary to expectations the considerable British investment in basic science was not reflected (at least over a long period) by the efficiency and competitive power of the British economy. Other limiting factors have prevented so far the full use of genuine knowledge and trained manpower.

The next step in my talk will bring us closer to the critical question. My thesis is that while most scientists agree that genuine knowledge is good in an ethical sense, the situation with respect to technology is totally different. Every technological achievement is *necessarily* ambivalent; that is, it can be good or bad, depending on one's point of view or on the particular situation. Technology is *necessarily* a double-edged tool. The classification of a given technological act or achievement as good or bad, right or wrong, is never certain because man is not omniscient. A given act can probably always be classified as ethically good or bad depending on the ends one has in mind, and depending on past, present, and future boundary conditions. The nuclear bombing of Hiroshima, which was intended to be and conceived of as an ethical act (to save the lives, both American and Japanese, which would be lost in a full-scale invasion), was later classified as unethical (2).

To put this in more abstract terms, propositional functions of the form "X is good" have no verifiable meaning until they are modified to the form "I predict that X will be found good for the purpose of attaining end Y under conditions Z." The end Y and the conditions Z must be inferred and supplied before the statement makes sense, that is, before it acquires verifiable meaning. The situation as it is can be summarized as follows:

1. Every true proposition, every piece of genuine knowledge, can be *potentially* applied in technology.
2. Every technological system or measure is always and necessarily ambivalent.
3. It is a grievous misunderstanding of science to interpret the inherent moral and factual ambivalence of technology as an ambivalence of genuine knowledge; and it is bitterly ironical that this belief is widespread among the younger generation.

This is a point I just do not understand since every real example you analyze will lead to the same conclusion: that the same piece of scientific knowledge can be used technologically to cure or to destroy, to erect a masterpiece of human culture or to destroy the beauty of nature. Modern medicine, or to be more specific, the successful application of the genuine knowledge of microbiology and immunobiology in biological technology, has decreased mortality at childbirth and during early childhood to nearly zero. This fact, which has been rightly praised as being one of the great achievements of modern medicine, has been by far the biggest contributor in large parts of the world to the population explosion that will inevitably destroy our civilization if not stopped very soon.

The discovery of nuclear fission by Hahn and Strassmann in 1938 led not only to the atom bomb but also to the peaceful use of atomic energy of which many nations, for instance my own country, are in desperate need in order to increase the supply of electrical power and to decrease pollution. The atom bomb itself is a terrible weapon, but the fears it causes may protect us against another big conventional war. I barely survived the last conventional war in Europe, and I do not hestitate to praise atom bombs if their potential for destruction prevents another war of this type and these dimensions. If ethics are not powerful enough to prevent war (and they never have been in the past), we cannot but prevent war by creating and maintaining strong fears. This is a bad solution, I agree, but it is much better than any type of conventional war.

In summarizing the statements I have made so far I shall return to the heart of the problem: the goal of science is knowledge, genuine knowledge, about the systems of the real world, including man. Absolute intellectual freedom is an indispensable prerequisite of science, and the scientific community must be allowed to pursue the search for truth according to its own ethical rules, the principles of which were established during the Renaissance and which have been obeyed since by every generation of scientists. The scientific community is responsible for the truth of scientific knowledge.

The goal of technology is to change the world according to decisions made by man. The data and laws and in particular the paradigms of science are the most important elements in modern technology. Our present world is totally adapted to and dependent on technology, and thus technology is an irreversible phenomenon. The regressive phenomena of technology that threaten human life (and the life of every other creature as well) can only be overcome by the use and considered application of *new* technology, which is again based on genuine scientific knowledge. Since many of these data are not yet available or not reliable enough to follow the formulation of laws, the progress of science is essential for our technological survival. Present-day scientific knowledge is clearly not satisfactory as a basis for future technology. This is especially true for the biological sciences and for the science of man. It is not only “mission-oriented” research that is urgently required (by this term I mean work done where the area of application is known); a similar emphasis must be applied to “curiosity-oriented” research (by this term I mean research done exclusively for the furtherance of genuine knowledge, thereby strengthening the intellectual framework of science and, consequently, the trustworthiness of the present-day paradigms on which any piece of technological innovation is at least indirectly dependent). Therefore, as I pointed out earlier, it would be a disaster for our own and every future generation if the rate of scientific progress were considerably reduced.

On the other hand, however, it is obvious that the technological application of scientific knowledge must be brought under rigorous control. Otherwise we shall destroy the surface of our little globe irreversibly, and this in the foreseeable future. What institution could exert this control? What is the function of science and what is the responsibility of the individual scientist and of the scientific community with regard to this critical problem?

Science can advise man how to attain a goal, but science cannot tell man whether he ought to choose that goal. In a free society that is necessarily pluralistic you will always find a multiplicity of goals, and for this reason any political decision must be a compromise and can never please everybody. Some people feel that there is one goal that is common to all people—survival of man as an individual and as a species. Although many arguments have been advanced in favor of this position, I doubt whether it can be derived from scientific knowledge. In any case, however, the assumption seems to be justified that the desire to survive as an individual and as a species is common to most people. If mankind were to accept this goal, which would require a political decision of gigantic dimensions, man would have to develop rapidly an appropriate moral system, that is, a pattern of individual and collective conduct that would maximize the probability of survival of man as individual and species in a world that is ruled inevitably by technology. I feel that the explicit formulation of the new ethics that must be obeyed to attain this goal would be mainly up to the scientific community, because the genuine knowledge of science must replace traditional ideologies and established ethical propositions, which are clearly outdated by the progress of our technologic culture. As an example, people must be told by science that it is a sin from the point of view of survival to have more than two or three children and that the only ethical behavior in this connection is rigorous birth control.

But let us return to reality: does mankind really exist? Is mankind an institution we can trust to the same extent that we can trust the scientific community? Have the United Nations ever been able to solve a problem? Clearly, the problem of controlling technology is an international, world-wide task; it is a problem of mankind, probably the first problem common to all people. But mankind is still a fiction; in political reality, mankind does not exist. There is, moreover, not the slightest chance that this situation can be improved in the foreseeable future, even under the pressure exerted by the threat of world-wide energy shortage and pollution, and in view of the warnings by eminent experts that it is not just the quality but the very survival of human civilization that is put in question. For this reason, we must design control systems for technology that work on the level of existing political units, that is, on the level of a state or a nation, hoping that these models can eventually be extrapolated to larger and politically more complex systems.

Before going on to describe such a model, I would like to confess frankly that I do not believe that our culture can survive for more than a brief period. The main reasons for my pessimism are the following:

1. Man is obviously not able to overcome his prescientific attitude of autistic thinking, as documented by the Bible, by which man made himself the center of the universe, created by God for his pleasure and temptation (9). In accordance with this prescientific attitude man became used to overpopulating, exploiting, and polluting the earth and to destroying step by step and irreversibly the genetic and the system potential of evolution. I am afraid that we shall not be able to modify our anthropocentric attitude within the short period of time which is all that remains available. Man has missed the right opportunity to overcome his prescientific approach to nature. Man should have changed his attitude toward nature when the new science-based technology made him the master of natural forces.

2. We observe an increase of violence, irrationalism, illusion, and daydreaming throughout the Western world just at the point when rationality is the main requirement for solving our problems. Instead of accepting the challenge and trying to master the problems of the modern world, a considerable part of the younger generation (at least in Western Europe) tends to ignore reality, including the reality of human nature and the reality of ecosystems, and indulges in dialectics and romanticism, sometimes combined with destructive criticism (which has always been the easy way). In the United States the "Counter Culture" as articulated by ROSZAK, MARCUSE, and GOODMAN is characterized by a radical critique of what they call technocracy and reductive rationality (10). However, behind the eloquent description of what people dislike and fear about technocracy, the Counter Culture is an open and sometimes very crude attack upon the intellectual discipline and the authority of science in favor of the unconscious, the irrational, and the mystical. In Continental Europe the "critical theory" advanced by some leading marxist intellectuals such as HABERMAS (11) has created considerable confusion among the academic youth and has caused a decline in several academic disciplines, so far in particular within the humanities. Fortunately the "new left" has tried in vain so far to find resonance in the public. While most scientists are likewise immune against the temptations of the neo-marxists, the art of clear thinking that is so characteristic of the scientific approach is being eroded rapidly as parts of our universities become infected by critical theory, which implies an antiscientific attitude, favoring Marx and Hegel instead.

It is sometimes stated, even here in the United States, that the university is the natural habitat of an honorable, in particular German radical tradition, the "critical tradition," that conceived the intellectuals' role as perennial opposition to any social order at all (12). While this might have been true for a number of years after the Second World War, it is certainly no longer true in present-day Europe. The critical tradition, as maintained after the Second World War by the "Frankfurt School", has become obliterated by the rising influence of devoted and sometimes stubborn marxists.

A characteristic of the present-day critical theory is to ignore scientific knowledge about nature as well as about the nature of man, to use nebulous terms and dialectic phrases and thus to escape from factual and logical criticism by the established sciences. What gave birth to science was a new way of thinking discovered in the Renaissance, the essence of which was the search for truth for truth's sake. Now, however, we are about to sacrifice the strength of scientific thinking in favor of a new kind of medieval theology just at the point in history when genuine, unbiased knowledge and the full intellectual potential of the contemporary generation is required more than ever to secure the future of mankind.

3. Is the scientific community, which is held together by the ethics of science, a political factor or even a power? The answer is clearly no. On political issues, by definition, there is no unanimity of viewpoint, and scientists are likely to be just as divided among themselves on such matters as is any "free" (i.e., potentially pluralistic) society at large (13). The scientific community exists only with respect to the furtherance of genuine knowledge. The ethics of science are only a rigorous guide as long as the elaboration and defense of objective knowledge is the goal. For this reason it seems unlikely that any kind of Hippocratic oath for scientists would get us very far. Every now and again there comes a time for such an oath, in which scientists should vow to work for some set of humanitarian ideals described in suitably vague terms, or to stop work if the goal is not in line with the ideals.

It amazed me to learn that DESCARTES at the end of the *Discours* uttered a kind of scientific oath. He said: "I could not work on projects that are useful to some only by being harmful to others." This is a statement of overwhelming simplicity that one would not expect from an intellectual hero.

A representative proposal advanced and widely distributed in 1970 by DULLAART (14) is:

> "Being admitted to the practice of the natural sciences I pledge to put my knowledge completely at the service of mankind. I shall prosecute my profession conscientiously and with dignity. I shall never collaborate in research aimed at the unjustified extermination of living organisms or the disturbance of the biological equilibrium which is harmful to mankind, neither shall I support such research in any way.

> Guide of my scientific work will be the promotion of the common welfare of mankind and in this context I shall not kill organisms nor shall I allow the killing of organisms for inferior, short-sighted, opportunistic reasons. I accept responsibility for unforeseen, harmful results directly originating from my work; I shall undo these results as far as lies in my power. This I vow voluntarily and on my word of honor."

There are some obvious difficulties (13):

1. Who will decide, in any particular case, what course of action will actually bring us closer to realizing the ideals? It has been the Hippocratic oath for physicians—the pledge to protect human life by all possible means—that has contributed to the deadly threat of mankind, the exponential population growth on an already crowded earth, and has contributed to the irreversible loss of animals, plants, and ecosystems on this planet. In brief: the Hippocratic oath has been an important factor in the reckless extinction of the wealth of genetic evolution.

2. The call for a Hippocratic oath arises from the realization that certain scientists are working to achieve some purpose of which some other group or individual disapproves. No oath, however, can eliminate *honest* differences of opinion and the legitimate pluralism of the scientific community on *political* issues of any kind.

3. It is an awkward truth that simple idealism and *good will* often turn out to be inadequate in the face of the complexities of the real world. A most serious problem arises if a scientist who is superior in his own discipline (e.g., a Nobel laureate) but unexperienced or even naive on political, philosophical, or ethical issues, raises his voice. There is the danger that the public and even the experienced politician will listen to him because they extrapolate from his competence in a particular field some competence in public affairs. The results can be terrible.

As an example, some of the wildest statements about the nature of man in the continuing battle about "nature and nurture", I heard from a Nobel Prize winner in the physical sciences.

4. In the thirties, marxist thinkers in Europe such as Bernal or Haldane openly used their high prestige as scientists to communicate and spread their ideological convictions. During the same period, some German physicists (among them 2 Nobel Prize laureates) created what they called German Physics, at least an indirect support of Nazi ideology. Let us assume these scientists were honorable men. What follows? Sincerity of purpose is no consolation if the actual effects are other than those intended. It is sometimes the do-gooders who do most harm. If the narrow-minded or ideology-infected do-gooders are renowned scientists, the worse! In any case, the scientific community cannot arrogate to themselves the right to make ultimate decisions.

After these admittedly personal and provocative remarks I want to return to my topic and to discuss the role of the *scientist* in making decisions as far as the development and application of technology are concerned.

The essence of decision-making in an open pluralistic society is to reconcile differing interests and to find the fairest compromise possible. "Interests" in this context do not include only materialistic motives but also nonmaterialistic values, irreducible ethics, and even individual or collective prejudices and faiths.

Traditionally, decisions are made at several levels by politicians (involving the legislature, military men, and administration) in accordance with the constitution of the particular country. Nowhere, to my knowledge, is the role of the scientist defined by the political and economic system, and correspondingly his responsibility is not clear. Politicians usually emphasize, rightly, that political decisions are based on experience, too, namely on the accumulated experience of a long cultural tradition and evolution. However, this experience is not free from logical or factual contradictions nor is it organized into a fully rational pattern. It involves emotional and instinctive elements as well as mystical articles of faith that have become a part of ideologies and even of constitutions. Moreover, it will include conscious and subconscious precepts for yielding power, satisfying ambition, and practicing demagogy.

Politicians are increasingly blamed by scientists and by technocrats for introducing as a matter of course these irrational elements into the course of decision-making. I feel that this criticism is not fully justified. Politics must *necessarily* involve irrational elements and irreducible values as long as we live in a free society and enjoy political pluralism, that is, have the right to choose our goal. The decision in favor of a particular goal, as a rule, involves a considerable degree of irrationalism, on the side of the citizen as well as on the side of the politician. I feel that this element should not be lost! The postulate that political freedom and pluralism should be sacrificed in order to make technocracy possible, can only be advanced by a person who does not know what dictatorship really means. A perfect dictatorship, that is, where all goals are set by one man or a few people, is a necessary prerequisite for a perfect technocracy. Therefore, I plead against technocracy and in favor of political freedom and pluralism, even if we have to make allowances for irrationalities in political decisions at all levels. However, irrationalism has become extremely dangerous in a world that is equipped with a tremendous and even deadly technological power. Obviously, political irrationalism must be tamed, even within the constitutional framework of a free society. How can this be achieved?

I have proposed on several occasions (15,16) the following model, which could be advanced in terms of a general systems theory but which can also be explained quite simply. The model has been developed for the situation in my own country, but it might be applicable, at least in principle, to every society that is based on personal freedom and on scientific technology. I firmly believe that the problems we are facing can only be tackled by a rigorously organized cooperation between scientists (including scientifically trained and oriented technologists) and decision-makers (that is, political representatives of the people), wherein the responsibilities are clearly defined. In this model, the decisions (including those about top research priorities) are to be made by the politicians, but they may only be made between alternative models that are elaborated, or at least approved, by scientists competent in the particular field in question. The scientist's responsibility is to ensure that only genuine knowledge is considered and that the ethical code of science is obeyed during the course of constructing the alternative models; this ensures, practically speaking, that every element and every relation in the model is reliable. On the other hand, the politician is responsible for the decision that is to be made between the different models. And at this point, irrationality (as pointed out previously) rightly comes into play.

Let me briefly describe the practical procedure of the two-step model of decision-making. Using different assumptions, different self-consistent systems can be constructed—as a rule expressed in form of "if–then propositions," each of which is equally logical and equally justified by scientific knowledge. The alternative systems (or models) constructed by the scientists (including scientifically minded technologists) are only different because different assumptions have been chosen. In most cases, the solution of a complicated problem in the real world can be based on different assumptions (depending on the means and values one has in mind), and a decision to use one assumption rather than another is required. In most cases, the scientist cannot make the decision because several sets of assumptions are equally justified from the point of view of science. In those cases, a decision based on political experience, political taste, political prejudice must necessarily come into play. An example (9): Ptolemy placed the earth at the center of the universe, Copernicus, the sun. Each system is perfectly logical and self-contained. Copernicus never ventured to give preference to his own system. The only way to decide between the two is to see where they lead if applied to a particular problem in the real world. We know now that an astronaut basing his calculations on Ptolemy would disappear into nowhere. However, at the time of Copernicus nobody could have made a fully rational decision between Ptolemy's and Copernicus' system if he had had the responsibility for a space program.

I would like to close my prologue by discussing a few of the critical points of the cooperative model of decision-making.

1. A very serious problem in this cooperative model is the manner of communication between scientists and decision-makers. Because there are hardly any scientists among the political representatives, the latter are not familiar with the manner in which problems are solved in science. Our standard liberal arts education does not teach them. Neither do the law or business schools. On the other hand most scientists have no real appreciation of the difficulties of politics; they are, in general at least, politically inept and commercially naive—they do not understand how to exert

political leadership, how to handle political power, to take over political responsibility, and to live with the necessity of continuous moral and factual compromises. My feeling is that the rigorous application of systems analysis and general systems theory might be the only means to a fruitful cooperation. The detailed knowledge of the elements of the system is up to the specialized scientist or technologist, whereas the analysis of the relations between the elements is predominantly a logical problem that can be followed and understood by any man with a high I.Q. if he is willing to learn the language of systems theory and to admit that nowadays political problems must be handled like logical systems and that any intuitive solution of a problem must be backed by logical solutions that take into account all assumptions, rational as well as irrational. Let me emphasize again that I do not advocate banning irrationality from politics but I do advocate the necessity of defining clearly where, that is, at which points of the system, irrational elements come into play.

It is not fair to blame the politicians for not being adjusted to the present situation. The scientific community in its ivory tower has not done much so far to insist on constant adjustment or to suggest means of improving this adjustment. The fact is that science and innovative technology have rapidly created a new world without allowing time for adjustment of the educational and political systems. This ever-increasing discrepancy is a latent but very serious danger for the perpetuation of democracy and political order.

The understanding of scientific propositions by the nonscientist is further complicated by the fact that each noun in science has a double reference (2). When the word "cell" is used, sometimes it refers to the conceptual model of the cell, other times it refers to the "thing-out-there," the postulated real system that the model attempts to represent. Without question, the semantic carelessness of scientists and the general lack of experience in the field of philosophy of science among scientists have considerably increased the difficulties of mutual understanding between scientists and nonscientists. This statement leads me directly to the second question:

2. Is a scientist able to cooperate with a politician at all? From my own experience I would say that by nature this cooperation is difficult; however, I firmly believe that it is possible if the duties and the responsibilities are clearly separated (as indicated in my model) and if both sides are really willing to cooperate toward achieving a common goal. One great problem I have already briefly touched upon has been the fact that many distinguished scientists (and even some less distinguished ones) are sometimes ambitious and conceited beyond the average, and thus they often believe that a high degree of professional competence and excellence in a particular field of science will automatically lead to a general competence, even in the field of politics and political morals. However, experience tells us that the political performances of scientists are amazingly poor, at least

in general. This is understandable: the scientist's structure of thinking and arguing is so different from the politician's structure of thinking and acting that the scientists (including Nobel Prize winners) must perform poorly in the political arena if confronted with an experienced and intelligent political professional who knows how to employ his manipulative skills. If an established scientist tries to turn into a political professional, e.g., as a member of a legislature or as a Minister of Science and Technology, he will rapidly lose his scientific competence and stature and as a rule he will fall (after a surprisingly short time) between two stools: neither the professional politicians nor his fellow scientists will take him seriously any more. Therefore, I plead for a cooperation in which both the scientists and the politicians remain professionals in their particular fields and do not try to ignore or alternatively to take over the specific duties and responsibilities of the other partner. At this point, I always express a serious warning. Scientists should renounce their attempts to impress politicians with super-optimistic statements about the logical order and the transparency of the system of science. It is neither fair nor harmless to evoke this impression. Despite some effort, the complex dynamics of the internal structures and external links of the system of science are not fully understood. Philosophy of science, sometimes optimistically called science of science, has by no means reached the status of a theory, free from internal contradictions and inconsistencies. We still do not understand many important feedbacks in the system of science, and thus the accuracy of predictions is usually very poor. We simply cannot predict with a high degree of probability how the system of science will respond if someone introduces a strong change. Therefore any short cuts and rash approximations on the part of the policy-maker regarding the system of science may lead to quite unexpected negative results. It is plain truth that any complex system that is obviously labile and that we do not understand sufficiently, must be treated with utmost caution to avoid an irreversible collapse. The system of science is such a system. Even though we do not like this fact, we may not ignore it.

3. Can the politician trust the scientist? Can he trust his models, his intellectual fairness? Can he trust his political neutrality? This is, in my experience, the most difficult problem, and the politician has the right and even the duty to be sceptical. Science has developed excellent methods for deciding whether a given proposition is true or false, and every experienced scientist knows that the intellectual and experimental methods in many scientific fields are amazingly trustworthy. These methods undoubtedly produce genuine knowledge, if handled rightly. Rightly in this case means adhering to established scientific behavior, as given in the behavior code for scientists, the ethics of science. This behavior code of science comprises a number of rigorous rules that must be obeyed in order to guarantee genuine knowledge: be honest, use only logic, never manipulate data, be fair, be without bias, do not make compromises, accept no authority except experience—these and a number of other rules are obeyed as a

matter of course by every scientist whenever he works as a scientist, i.e., whenever he is concerned with obtaining or processing data. It has always amazed me to watch scientists (including myself) carefully and to notice how fast even those people who tend to be extremely biased, unreliable, and even untruthful in their private lives will change their attitude as soon as they approach the workbench or the desk to do *scientific* work. Complete honesty is of course imperative in scientific work. Without discussing the ethical code of science any further at this point, I would just like to say that in the long run it pays the scientist to be absolutely honest, not only by not making false statements but by giving full expression to facts that are opposed to his views. Moral slovenliness is visited with far severer penalties in the scientific than in the business or political worlds (17). In this context, an oath is not only realistic but even necessary and has long been introduced at some of the European universities. At my own institution (in Freiburg) a text of the oath is still in use that was originally introduced in Königsberg by Immanuel Kant, the famous philosopher of the Age of Enlightenment. The language sounds somewhat old-fashioned but the essence of the oath is perfectly up to date:

> "The faculty has resolved to confer upon you the degree of doctor of science. It is an honor which carries with it an obligation. The obligation is forever to remain loyal to scientific truth, never to yield to the temptation to suppress or to falsify this truth, whether under economic or under political duress. In this spirit I, as dean of the faculty, offer you my hand and pledge you to preserve from every stain the honor which the faculty now confers upon you, and uninfluenced by other considerations, to seek, and to give your allegiance to, only the truth".

A serious violation of the oath if proved will lead to the annulment of the doctor's title. This has been a rare event, of course, but in principle the faculty would not hesitate even today to draw these consequences.

In order to avoid any misunderstanding, I would like to add the following restrictive remarks: scientists as a rule are not exceptionally ethical or modest in their personal lives, and there is no reason to expect them to behave better than or even differently from other people. Some scientists even behave badly toward their fellow scientists as soon as competition, priority, prestige, and sometimes even money come into play. While these deficiencies are not particularly liked, they will eventually be ignored or tolerated by the scientific community, if the scientist in question does excellent work and has never violated his real reputation, which is for being absolutely honest and trustworthy in scientific matters, that is, in obtaining and handling data and in using logic.

To summarize: a scientist does not have to fit into the same categories in his work and in his life, but if he does not, he must be content to exist as a dual personality. And he must be aware of this fact.

Against this background we again ask the question of whether the politician can really trust the scientific community to be both absolutely competent and absolutely honest in *scientific* matters. I have the unhappy feeling that at present the answer is no. This is a very serious verdict and I need a few minutes to justify my statement. The problem involves at least two aspects: complexity and ideology.

Let me deal first with the aspect of complexity. We all realize, I guess, that certain fields of science, which are concerned with very complex systems, are still in their infancy. In these fields, the formulation of general propositions, laws, or law-like statements, is still a risk, if in fact it is possible at all. Even some important fields of biology must be placed in this category, including environmental sciences, psychology, and the science of behavior. Let me briefly describe the principal situation in these fields, using the science of economics, a field of tremendous complexity, as an example (2). The science of economics appears to be at a stage of development analogous to that of physics at the time of Galileo, and it seems premature to try to construct a general economic model at this stage. This is not only true for the economy in the western world, where not only multiple demands but also multiple ideologies and convictions as expressed by government, free enterprise, labor unions, and pressure groups of all kinds contribute in a nearly unpredictable manner to the multifactorial economic system. In those communistic countries where the economy is based on general models derived in essence from the political philosophy, the results so far have been very poor and sometimes even disastrous for the people. It remains to be seen how China will perform as soon as a minimum level of prosperity has been established and the desire of the people for individuality, diversity, and a limited economic freedom becomes inevitable. In any case, there is no general economic model that would satisfy the two requirements of being completely rational and at the same time satisfactory. However, it is possible to isolate certain aspects of economics and to attempt to construct rigorous models of these isolated systems, models that can be checked scientifically, that is, by prediction and experience. This "operations research" has for some time been regarded as a positive example of cooperation between science and government. While the reputation of some operations research groups has been marred by their predominant involvement in war planning, operations research as a whole has remained an outstanding example of fruitful cooperation between scientific experts and nonscientific decision-makers, even in one of the most complex and most complicated fields of science.

The severe problems we are facing in the large cities throughout the world are, as a rule, not due to a failure of operations research (even though the paucity of numerical data and the dependency on observation rather than experiments are serious handicaps); it is the political aspect of the problem that prevents a rational and straightforward solution.

Economic systems are particularly difficult to study even by the methods of operations research because these systems are not only "open" ones but are also susceptible to outside and psychological disturbances. As an example, it has been very disappointing for a systems analysis fan like myself to observe that even those models of operations research that are characterized by a high degree of inner perfection and certainly deserve at least some confidence have virtually failed in fighting the economic difficulties of Western Europe. The reason for this is that new variables come into play that could not be accommodated by the models, in particular a dramatic and rapid change in the climate of public opinion and confidence that led to an unprecedented decline of incentive and investment. I do not want to speculate about the causative factors for this loss of confidence on the part of the public, but this is related to the second aspect I would like to consider briefly, namely, the impact of science and ideology on our present European universities. A number of fields at our universities have been using the term science without deserving it, and as a consequence the formerly sharp distinction between speculative and scientific statements has rapidly disappeared. The high rank and reputation of the classical scientific fields are misused by these disciplines to support their own non- or semiscientific statements. This is especially true for some fashionable disciplines within the social and political "sciences," but it is even true for parts of psychology and education. Many models developed in these fields contain more ideology than genuine knowledge about the nature of man and civilization. Therefore, the models do not work if accepted by the politician and applied to the real world. As an unhappy example from my own country, I could mention some current models in the field of education, including the structure of our newly-founded universities. These repeated failures have diminished confidence in the reliability of science as a whole, among the public as well as among those politicians who are aware of their duties and responsibilities. It will be very hard to repair the damage already done to the reputation of science, just at a time when the full power of the true sciences is required more than ever to develop reliable models that could be used to overcome the terrific problems of our age. In any event, at present, the bewildered citizen as well as the politician no longer has any reason to place blind trust in scientific models. Rather he must take into account that they might be biased by ideology and even personal prejudice.

It is absolutely legitimate that a scientist is guided in his personal and political life by collective ideologies or individual prejudice, including articles of faith and irreducible values. However, as soon as we act *as scientists* or contribute to a topic of public interest explicitly as scientists or even as scientific experts and advisers, we must renounce ideology and articles of faith. It is a sign of sincere corruption if a scientist enters the political arena to use (or rather to misuse) intentionally or carelessly his prestige in support of a particular private or collective prejudice. This kind of behavior has been detrimental to the prestige of science in the public as well as among conscientious politicians throughout the Western world. The depressing outcome is that the problem in question is spoiled for a serious and reliable scientific analysis because, in the eyes of the public, science has lost trustworthiness in dealing with this particular matter. The Teller–Pauling confrontation has had this consequence in the past; the Jensen–Hirsch type of confrontation, observable at present (18,19), will possibly stir up the public still more and lead to a serious, long-term loss of the reputation of the science of man.

I shall close my prologue by summarizing the main points:

1. There are certain minimum requirements that a scientist must fulfill in order to be able to advise the politician within the framework of any cooperative model. The scientist must have a solid knowledge about the solid facts of a solid discipline. He must have sufficient knowledge in systems analysis or general systems theory. Also, he must have an innate drive to improve the social system. All these requirements must be met. A strong motivation by itself is not sufficient. A scientist who meets these requirements must therefore prevent the participation of others who have a strong desire to improve the world in the interests of science without having the capability of logically analyzing complicated situations and without being knowledgeable and experienced scientists.

2. The responsibilities of any member of the scientific community can be summarized as follows: we must preserve and rapidly increase the amount of genuine knowledge. We must reestablish the full trustworthiness of scientific propositions by eliminating completely the participation of those people and fields that do not rigorously obey the basic rules of the ethical code of science. We must contribute deliberately to the rationality of political decisions by thinking in terms of general systems theory and by advising the politician, in accordance with the cooperative model, even at the expense of our scientific career in a particular field of research. Science and science-based technology is man's only means of survival. If we allow this marvelous instrument to be damaged, or if we do not apply it rightly, man will disappear from this planet.

As a kind of appendix, I would like to say a few words at this point about the function and responsibility of the university, in particular with regard to the so-called "Politicized University" that has developed in Europe during the past eight years (12). The typical German university, as designed in principle by Humboldt at the beginning of the 19th century, has been an explicitly apolitical republic of scholars. It is dedicated to the student's right to learn and to the professor's right to inquire and to teach in unlimited academic freedom. Humboldt's republic of scholars was deliberately conceived as a revolutionary institution to the extent that pursuit of genuine knowledge, irrespective of princely whims, approval by religious authority, or social utility, was a revolutionary notion. The German university, in the tradition of Humboldt's reform, had by the end of the 19th century become a very strong institution, in fact strong enough to survive Bismark, Wilhelm II (20), the First World War, and even Hitler, attenuated but largely intact. Humboldt's model had left its mark not only on German universities, but also on everyone else's. The idea of a modern university, independent of ideological, religious, and even economic traditions, and devoted exclusively to the pursuit of truth, irrespective of national boundaries (20), has influenced the systems of higher education all over the world. Humboldt's university was strong enough to contribute essentially to the restoration of West Germany and Western Europe after the Second World War. However, it decayed rapidly after the compulsory politicization of academic affairs beginning about 1967, following the hasty installation of the new group university by an inconsiderate and irresponsible legislature. Maybe the traditional Humboldt university did not pay enough attention to the social context and to modern, up-to-date governance. The Humboldt university was predominantly based on the excellence of the leading professors. Unfortunately most professors with high scientific reputations are poor politicians, in particular bad campus politicians. The motto: truth first, relevance second, was perhaps too rigorously maintained in a changing social climate with strong demands for actual know-how. On the other hand, it has been the strength of the Humboldt university to follow its own rules, to build up the paradigms of an age, and thus to serve best the long-term requirements of an open, pluralistic society. The Humboldt university has always refused—in principle even under Hitler—to act in favor of any particular political ideology or power.

The group university, which was supposed to reflect in academia the spectrum of political convictions in society, was built on myth rather than on solid ideas about the function of a university in a modern society. The reactions of the German universities differed widely. Some of them, in particular those with strong science departments, have returned, after a short phase of turmoil, to the spirit of scientific inquiry and teaching; others became politicized universities in which the left (sometimes dogmatic marxists) prevails, which does not at all reflect the distribution of political forces in our society. The revolutionary tendencies manifested by some students and professional intellectuals seem hardly to touch the real working classes. However, for some universities the development has become deleterious. The politicized universities have been used, quite openly, for revolutionary rather than for academic purposes. The dedication to scholarship and the pursuit of truth are being ridiculed in favor of vague and premature "project research," which is bound to fail since it lacks seriousness, consistency, and clear objectives as well as scientific competence.

In my opinion, the university's general responsibility toward the society that supports it, is undeniable. On the other hand, this responsibility can only be fulfilled if the university is not only free from direct political pressure and manipulation but is permitted to maintain some *distance* from the political and social system in which it is embedded. *Distance*, also *emotional* distance, from the system I want to understand, has always been a prerequisite of the scientific endeavor. The violation of this principle has led to the politicized university in which a professor must be political all the time and totally preoccupied with defense and counterattack. This kind of a university is no longer a favorable place for the pursuit of truth.

A politicized university is neither a place for scholarship nor a place for sincere teaching and learning. Worse, it is by no means a place to educate students for their professions *and* for the assumption of responsibility in an open, democratic and necessarily pluralistic society. It is the *depoliticized* university with a high scientific and moral standard that educates the young generation to take over political responsibility in a free society.

In conclusion, the antirational and particularly antiscientific attitude of the counterculture or of the politicized university is obsolete and without future. We simply cannot afford, for practical reasons, this kind of behavior. On the other hand, however, science cannot derive values to cover the whole of human existence. The ethical code of science is restricted to the realm of science: the ethical code of science is only concerned with the finding and preserving of genuine knowledge. This implies that the sense of science cannot be found inside the sciences. The justification of science is a matter of philosophy rather than a matter of science. Science is not self-sufficient with regard to the justification of its own existence.

2nd Lecture

The Motivation of Science

It is obvious that this problem has two different aspects: (1) the motivation of society to support science and (2) the motivation of individuals to become members of the scientific community. The first aspect must be understood before the second aspect can be envisaged (1).

The Motivation of Society. It is honest and necessary to discriminate at any stage of the discussion between science and technology (industry, commerce) rather than lumping the two together. Science costs money, manpower, and material investments; it is technology (after appropriate innovation) that makes money and increases the wealth of the people (see Fig. 22). Most people in Europe or North America will appreciate (vaguely) the cultural values of scientific research. Even after the rise of modern technology (in the second half of the 19th century) "scientific culture" was considered within the educated classes of the leading European nations to be "independent of the applications of science, for it was dedicated to a knowledge of nature, man, and humanity—a knowledge desired for its own sake—and to a world view based on that knowledge" (10). However, what most people wanted from science has been in general wealth and security, freedom from hunger, freedom from poverty and from disease, and freedom from military threat. At present, there can be no doubt that the presumed close connection between science and the material welfare of a country (as expressed in terms of the standard of living, increase of life expectancy, or military and economic power) is the main reason why the public and governments support science. If the public ceased to believe in the link between science and material welfare, support for science would rapidly diminish. Why?

The pursuit of knowledge for its own sake is the noble ideal for a scientist: knowledge for the sake of understanding, not merely to prevail, this is the essence of the scientific approach. However, is the normal citizen (Mr. and Mrs. Smith) really interested in the composition of the interstellar matter or in the molecular nature of the lac-repressor? If they are not, or not particularly, why do they nevertheless support science, explicitly or indirectly? For practical reasons! For reasons of survival!

The existence of science is not a matter of course. During long periods of man's cultural evolution anything comparable to modern science did not exist. Modern science started very slowly approximately 400 years ago and has since developed according to an exponential function. The strong moral and material support of science by the public is a very recent event. In CHARGAFF'S (2) words: "No one who entered science within the past 30 years or so can imagine how small the scientific establishment then was. The selection process operated mainly through a form of an initial vow of poverty. Apart from industrial employment, important for a few scientific disciplines, such as chemistry, there were few university posts, and they were mostly ill-paid. One of my former chiefs at that time assured me that for him the opportunity of doing research as he pleased was sufficient recompense. (He had, besides, a comfortable private income)."

We have rapidly become accustomed to the splendid situation of today, but it would be frivolous to assume that this favorable situation will last as a matter of course if the public becomes disappointed in the performance, efficiency, and reliability of science. The motivation of society to support science is predominantly a secondary motivation and thus unstable, reversible, and subject to decay as soon as the primary motivation, strengthening the prerequisites of civilization and power with the aid of science, loses ground.

Why is this necessarily so? The appreciation of objective knowledge for the sake of knowledge is a product of a cultural evolution and thus labile and potentially subject to decay at any time. By nature, as an outcome of genetic evolution, we are only interested in genuine knowledge for the sake of survival. In the course of the genetic evolution of man, there has been no selection value for genuine knowledge without application.

Even Francis Bacon (the masterly promotor of the scientific method in the early decades of the 17th century) emphasized that genuine knowledge was useful because it gives man sovereignity over nature ("knowledge is power").

The objection—not uncommon in our days of tight budgets—that basic science is in fact an expensive luxury can easily be invalidated. As we will consider later (see 12th Lecture; Science and Technology), most of our present-day technology, including medicine, agriculture, and landscape restoration, is based on the facts, laws, and in particular on the paradigms of basic science. On the other hand, according to WEISSKOPF (3), the total outlay for basic science—from Aristotle to Sabin—has cost no more than ten days of production in the United States. Is the scientific enterprise then so expensive? But nevertheless the climate has definitely changed. As

KETELAAR pointed out (9): "we have come from an age where the individual professor could more or less do what he wanted to do for a limited sum of money, but that time has passed. The limitation now is not science, and not curiosity, but the total amount of available money and labor, for all the things we want to do. And it will always be the case that for each project there will be only a very small group relative to the total population that is fundamentally interested. The degree to which they have to convince the others of the worthiness of their project is a matter of education."

It is, however, another aspect that deserves our deep concern, since it is becoming increasingly important these days. Objective knowledge about the real world is not only a tremendous advantage if we want to deal with the real world; objective knowledge is also a threat. It is, potentially at least, a risk and a menace for any kind of prejudice, for prescientific convictions, dogma, and articles of faith. All men are guided in their life, to some extent, by prescientific and irrational convictions, and most people greatly dislike giving up those helpful and beloved prejudices and individual convictions (which are mostly an integral part of our psychic homeostasis) for the sake of objective knowledge.

This is the reason for the annoying dialectic behavior of most people even in the industrialized and intellectually advanced countries in Western Europe or North America. At the very least, most people do not hesitate to enjoy the material fruits of science, wealth and security, freedom from hunger, from poverty, and from disease. However, they are also very reluctant to acknowledge objective knowledge as a superior authority in their personal or group lives if the objective knowledge contradicts their prejudice. There have been many instances of this kind of "dialectic" behavior even in recent decades, ranging from the embittered resistance of religious groups against Darwinian biology to the equally bitter present-day battle about the genetic determination of human traits, nature vs. nurture (4).

Every experienced biologist knows that measures of individual differences in almost any characteristic, and especially in ability, render the assertion of our founding fathers that "all men are created equal" patently false. While the notion of equality is consistent with biological knowledge when applied in two areas–(1) that of opportunity, in the sense that a society be without castes, and (2) before the law, strict egalitarianism—the doctrine that insists upon an equal distribution of at least all primary goods, can no longer be justified by biological knowledge about the nature of man. Rather, this doctrine is based on vulgar envy (which should not be confused with the sense of justice and dignity). Can we really assume that the public remains interested in objective knowledge about the nature of man under all circumstances? Could one expect the Christian churches to support the progress of the theory of evolution after the appearance of Darwin's *Origin of Species* and *Descent of Man*?

In those countries that are ruled by a rigid state philosophy such as marxism or dialectic materialism, the situation is in fact precarious. Objective knowledge, including the prevailing paradigms of science and the cultivated way of scientific thinking, is urgently required as a basis for the advance of physical and biological technology. On the other hand, however, objective knowledge (in particular objective knowledge about the nature of man) is hated by the ruling ideologists as it is the most serious menace of the ideological construction of Marxism (5).

The Motivation of the Individual. What motivates an individual to become a scientist and therewith to become subject to the rigid rules that govern the scientific community? To be a scientist means nowadays to have a respectable profession that provides enough means to live and to maintain a family. Looked at in this way science is not very different from other academic professions; but it is obvious that in general the motivation of a scientist is more complex and consists of several factors. Scientists as a rule are not seeking power for the sake of power (although they are not immune to the temptations of power); scientists as a rule are not seeking fortune for the sake of fortune (although they are not immune to the temptation of becoming wealthy); first of all, they seek recognition. Certainly, the primary motivation to contribute to the furtherance of genuine knowledge is not fiction or simply a product of wishful thinking about the "nature of scientists." I have met many people in my life for whom I believe it was only the deep interest in how nature works that motivated them to become scientists. Some of them have worked under adverse conditions and painfully struggled—as our scientific ancestors did—their way through academic or research institutions, neglecting income, family, administrative positions, and formal prestige. However, eventually they were all extremely interested in recognition. HARRIS (6) quotes an interesting remark made by BERTRAND RUSSELL: "One can't think hard from a mere sense of duty. I need little successes from time to time to keep . . . a source of energy." This latter observation seems to characterize all notable and successful scientists. Scientists (as a rule) are very ambitious. They want to publish ("see their name and their ideas printed"); they want to gain esteem and recognition from their fellow scientists (preferably from the higher ranks of the hierarchically structured scientific community); they fight fiercely for priority of their discoveries if necessary. It must be added that envy (mostly in a cultivated form) and sometimes even hate are not uncommon within the scientific community. Taken together these observations support the idea [developed in detail by HAGSTROM and JEVONS (1,7)] that recognition from fellow scientists is a major motive for a scientist, a permanent drive to work hard, not to violate the ethics of science, to be creative and, if possible by any means, to be the first with a new discovery. As LEOPOLD (8) puts it, "The notion of a career in science as an "ivory tower" existence of insulated introspection cannot be correct." Rather, it is because scientists desire recognition that they adopt the ethical norms of the scientific community. The belief (sometimes cultivated by scientists as well as by nonscientists who admire science) that scientists are driven exclusively by the desire to contribute anonymously to the advancement of knowledge is simply not correct in view of what really is going on in laboratories, in the offices of editors, and in formal or informal discussions. Is there anything wrong about this? I think not. Rather, I find it fascinating that in the course of cultural evolution the striving for recognition (more or less characteristic for most human beings) has been cultivated by the scientific community to an extent that the most valuable treasure of cultural evolution (besides art and poetry) became accumulated; genuine knowledge about the real world, including man. It is by no means a matter of course that recognition is generally regarded by the scientist as the highest reward, far more important than money or any official position.

The function of the scientific community (7) is two-fold in this connection: to grant recognition to the individual scientist and to control the activity and the behavior of the scientist from the point of view of the norms of the scientific community. This includes punishments (if a scientist has violated the norms) from withholding recognition to expelling an individual from the scientific community. We will return to this point later in the lecture on the Ethics of Science.

My strong emphasis on the reward system of the scientific community should not evoke the impression that I underestimate or deny the *personal* satisfaction in achieving a goal. The solution of a serious problem in science is always accompanied by intellectual reward. However, since this feeling is not *specific* for the scientist, it must not be considered in detail when discussing the motivation of the scientist.

The intrinsic reward system of the scientific community is sometimes corrupt or at least defective. Justice is a difficult postulate in science as it is elsewhere. There are scientists whose reputations are too high compared to their merits (so-called "over-rated scientists") and there are others who receive no decent recognition because some dominating person in their particular field (called the "pope") dislikes them for some reason or other. Editors and reviewers of renowned scientific journals who have become tired and reluctant toward new discoveries, new terms, new ideas, can easily discourage creative (young) scientists and postpone or even prevent their recognition. One of the traumatic experiences of my own academic career was the breakdown of a young scientist whose first paper (the essence of his thesis) was declined with excessively insulting arguments by the editor of a renowned journal. I still remember phrases such as "fairy tales instead of solid science," etc. The fact that the fascinating discoveries of the student are nowadays acknowledged by the experts in the field shows that discoveries can indeed be postponed by the misbehavior of "popes" but not prevented. Unfortunately, for the student the recognition came too late. He had irreversibly quit science after the debacle of his thesis.

To develop this observation in general terms: among those students who have chosen science under the drive of a primary, mainly irrational motivation, much of the original choice of career has depended upon high hopes and aspirations. At least some of these people had to realize that their phantasies clashed rather sharply with reality. Indeed, the degree of disillusionment inside a closed profession like science is generally high. This is not only true with regard to the modest material reward; the harsh treatment a young scientist sometimes receives from the establishment of the scientific community is much more important and often leads to traumatic experiences that are difficult to overcome later.

These are deficiencies that can hardly be avoided. In general, the delicate reward/punishment system, intrinsic to the scientific community, works not only very efficiently but also rather fairly. To maintain a strong sense of fairness is not an easy task. It is felt, sometimes, that some prominent people earn too much reward for their achievements. Indeed, the analysis of individual discoveries has shown that, in practically every case, the individual acted as a catalyst for a complex process in which many other individuals played a role. This does not diminish, in my opinion, the credit to be given to an individual for having advanced a field in a particular way; it does mean, however, that eminent leadership and even scientific genius is only one factor in the multifactorial process of scientific progress.

From the many complexities that could be discussed, I will only touch upon two important cases: extrinsic rewards and recognition in the age of teamwork (1,7). It would show ignorance of human nature to assume that "extrinsic rewards" (extrinsic with regard to the scientific community) are totally meaningless. However, I agree with JEVON'S and HAGSTROM'S analysis that recognition (an intrinsic reward) comes first and extrinsic rewards (position, money) follow from recognition. In practice, one can easily detect deviations from the ideal sequence. Most of these defects are explained in JEVON'S book (1) by the "Matthew Effect" ("unto everyone that hath shall be given"). As an example, a scientist who already holds a good appointment is in a particularly good position when a committee considers him for a higher one.

The second point, recognition in the age of teamwork, is an essential one because the motivation of scientists to cooperate depends on a fair solution of the psychological problems involved in teamwork. The recognition of individual scientists by the community of scientific colleagues is the classical case; the recognition of teams (a kind of shared recognition) is much more difficult because the "Matthew Effect" becomes effective (the most distinguished author, in general the head of the research unit, will automatically earn most of the credit of a paper even though his name is last in the list of authors or he is mentioned only in the Acknowledgements). On the other hand, the paper of a young, so far unknown author has a much higher probability of being read and recognized if it is co-authored by a scientist with a high status. Thus, as a matter of fact, the shared recognition within a team will in general promote a young scientist, at least if the team is really good and if the young scientist is given the chance to give lectures and to address the community of scientific colleagues during congresses.

The recognition of institutions is already common in certain fields, e.g., the big particle accelerators such as CERN. It is characteristic that the work of these institutions depends on the close, integrated cooperation of a large number of people. It is difficult to apply the classical recognition model to such cases because these institutions have to some extent escaped the control by the rest of the scientific community. They have become self-sustaining insofar as they have developed internal control mechanisms and a strong internal hierarchy. In addition an "extrinsic" control is exerted by governmental agencies. The institutions have become used to responding to this extrinsic reward/punishment system rather than to the intrinsic reward/punishment system of the scientific community, in particular if the self-esteem of the institution is very high (e.g., stabilized by the presence of Nobel prize laureates).

For the individual scientist, in particular for the academic scientist, two kinds of recognition must be kept apart: cosmopolitan recognition and local recognition. Cosmopolitan refers to recognition by the world-wide scientific community; local refers to the recognition the particular scientist receives from the other competent members of his institution (e.g., his university) even though these local colleagues do not belong to the same discipline. There can be no doubt that in many cases a lack of cosmopolitan recognition can be compensated for by local recognition and vice versa. The local recognition can also consider those merits of the particular scientst that cannot be appreciated by the scattered community of his professional colleagues, such as successful teaching, skillful administration, authorship of textbooks, popularization, and (in precarious cases) defense of science.

In the late sixties and early seventies, there has been a swing from science (in particular from physical sciences) in most of the Western world. This has been correlated with a rise of irrationalism (antiscience movements), naive political ideologies, and illusionary educational systems, in particular in Western Europe. Whatever the reasons are, we can describe this swing from science in terms of a decreasing motivation for science, in the public as well as among potential students of the sciences. This is a positively dangerous situation that must be counteracted. How?

There have been some proposals to increase the *secondary* motivations of future scientists, e.g., make the graduate studies, in particular the Ph.D. studies, more attractive financially; establish privileges for science teachers comparable to other fields (more pay and better promotion); grant somewhat higher stipends for science undergraduates as compared to other fields. As an incentive to young students to work in fields that the government judges to be of priority, up to 1500 studentships were created by the French government in late fall 1975, paying 2000 francs per month for the second and third years of postgraduate thesis work.

I feel that these measures will not work satisfactorily if we are not able to stimulate the *primary* motivation for science. By this I simply mean that our primary and secondary schools, the mass media, and capable authors of popular books must do their best to elicit in the young generation a deep interest in how nature works and to make clear what it means to be a member of the scientific community, to be subject to the ethical code of science, and to earn recognition for genuine discovery.

More recently, there seems to be a tendency back to science, at least in the United States. Some observers believe that this is not only due to improved professional opportunities but also to an increase of primary motivation for science among students. If this tendency indeed not only reflects opportunism but coincides with an increasing consciousness of the young generation for the nature of science and for the responsibility of the scientist, this would be a hopeful sign.

3rd Lecture

The Scientific Approach (1): Terminology and Language

First of all a sceptical remark: we are going to consider some aspects of the scientific approach. What can we expect from this kind of analysis? Can we expect normative insights? Or will our effort lead at best to a post factum formal reconstruction of the scientific way of thinking? Can we become more conscious, more reliable, and thus better and even more successful scientists? Or is it only that we try to make explicit what every experienced scientist "knows" anyhow? Is it really worthwhile for a practicing scientist to reflect about the foundation of science, or is it better to take these foundations for granted and simply use them unconsciously and intuitively? This is the alternative that has been preferred by most scientists so far. As HOLTON (1) noticed: "How do research scientists go about obtaining knowledge? How *should* they? Today's scientists tend not to be introspective about these questions. During their apprenticeship, they somehow absorb the necessary pragmatic basis and then go about their business, content to leave it to a very few among them to interest themselves in epistemology when some obstinate difficulty blocks advance." Nevertheless science has advanced rapidly (whatever "advance" might mean in this context) and the fact that a tremendous wealth of genuine knowledge has been accumulated by scientists who did not pay much attention to the theoretical foundations of science is beyond question. However, this theoretical abstinence obviously fails if a scientific field reaches maturity (e.g., physics at the arrival of quantum mechanics), as indicated by the loss of visualization, the restriction to probability statements, and concomitant *unavoidable* difficulties with terms such as "determination," "causality," and "prediction." Since biology is approaching a comparable level of sophistication in several fields, such as population genetics, molecular biology, and parts of system-oriented physiology, it might be advisable to consider at least in principle the theoretical foundation and justification (if any) of what we are actually doing whenever we practice science.

To quote JOHN ECCLES (1a), fabulous neuroscientist and Nobel Prize winner: "I can state that it is of the utmost importance for scientists, and particularly for the leaders of scientific research, to be illuminated and guided by a *theory* of scientific method," and HOLTON (1) warns that for the present-day scientist "this lack of attention has become a costly luxury, that a whole range of problems—from questions about the validity of the scientific components of public policy decisions to the need for a better understanding of the roots of public conceptions and misconceptions about science—would be clarified by a more widespread apprehension of how scientific ideas are in fact obtained and tested."

Science is a social phenomenon in the sense that it is a matter of the scientific community. Scientific knowledge is public knowledge (2). This implies that science is interested in clear and precise statements (propositions, judgements) that are supposedly "true" (i.e., in accordance with nature) and can be communicated. Some theoreticians of science [e.g., NORMAN CAMPBELL (3) define science as the study of those judgements concerning which universal agreement (among competent scientists) can be obtained, and WOODGER (4) states that science can only deal with shared thoughts, and thoughts can only be shared effectively by the use of language. WOODGER defines biology as "the system of accepted written or printed biological statements." Thus, the very essence of science is communication between competent people. The means of communication is an appropriate scientific language.

Languages are the main means of communication in man. The multiplicity and diversity of natural languages show that there have been many different trials in the genetic and cultural evolution of man to develop and optimize this system of communication. Terminology and the implicit and explicit rules of any particular natural language (e.g., grammar) must be looked at from the point of view of practical advantage for a human population under the particular conditions of life in the course of a genetic evolution. While natural languages must have had some strong selection advantage for the particular population, they are obviously not fully adequate for a scientific description of nature. Natural languages and even the system of natural logic to which we all were introduced in our childhood are generally considered nowadays to be inadequate for science. It was indeed the introduction of symbolic languages that has strongly promoted the development of science, in particular in physics and chemistry. The symbolic language that offers the highest degree of precision and universality is mathematics. It is generally agreed that the use of mathematics is the non plus ultra in formulation and communication of scientific propositions.

Is there any justification for the preference given to a particular symbolic language as being infallible in the description of nature? Why is it legitimate to apply to a wheel (a physical object) the mathematical formula derived for a circle

$$c = 2\pi r$$

or to describe the variability within a population of biological things with the same formula GAUSS developed for the purely mathematical probability distribution? GALILEO stated in 1623 that "nature is written in mathematical language" (and this phrase has been repeated by many proud, "exact" scientists over and over), but he could not give any explanation why this is so. In brief: why is mathematics applicable to nature? This problem is an aspect of a wider problem, the justification for formal logic, which is assumed to be applicable to every scientific proposition. Why? The justification of logic (considered here to be theory of deductive argument) is one of the most difficult problems that exists. The question has been, are the logical axioms necessary in the sense of an inherent law of nature or do they only possess a kind of psychological necessity or are they based simply on convention? (5). In any case, we may assume that the axioms and laws of logic (including the axioms and theorems of mathematics) have been created and formulated by man. The question arises, why did the human brain select these particular axioms and why does every intelligent human being (at least in science) use the logical laws and mathematical theorems as a matter of course. The answer to this question comes from science—from the theory of evolution—not from philosophical speculation and not from logic itself: man is a product of genetic evolution. Our mental ability to discover the structure of nature has developed in the course of genetic evolution. The intellectual capacity of man has been an adaptive trait of tremendous selective advantage. The gradual coincidence of the categories of nature and the (genetically determined) categories of the human mind has been *the* selective advantage that made possible the rapid ascent of man in the course of genetic evolution. In other words: correct thinking (i.e., thinking in accordance with the categories of nature) has been the strongest selective advantage in the course of genetic evolution; wrong thinking (i.e., thinking in discordance with the categories of nature) has been a strong selective disadvantage in the course of genetic evolution. A human being who was genetically enabled to make a "true theory" about some relevant aspect of nature could perform reliable predictions and thus develop technology; a human being whose genetically determined categories of thinking led to a more or less "wrong theory" about some relevant aspects of nature could not perform reliable predictions and thus necessarily failed with respect to reproductive fitness under the conditions of natural evolution. "Wrong thinking" and thus wrong action must have been much worse than no thinking (which implies dependence on strictly genetically determined instinctive behavior and unreflected learning).

After thinking and teleological action was "invented" in the course of evolution, a strong selective pressure in favor of "right thinking" has been maintained during further genetic evolution. The result has been a far-reaching coincidence of the structure of nature and the structure of thinking.

This is an example of the explanation of a problem of philosophy by a scientific theory. This is not a vicious circle as some traditional philosophers might object. Rather it is an example of the fruitful and indispensable cooperation and interaction between science and philosophy of science, in particular between science and epistemology. While this kind of positive interaction between science and metascience is more and more supported by scientists, many philosophers, in particular in Europe, are quite reluctant. This is unwise, because the consequence will be that experienced scientists with a genuine interest in the philosophy of science will gradually take over this discipline.

We return to our proper problem, justification of logic by information that transcends logic. The experience that axiomatic systems cannot be justified completely without additional information from outside the system is not new. The logician KURT GÖDEL showed in 1931 that all axiomatic systems (except very simple ones) must contain assertions that can be neither proved nor disproved by logical deduction with the given axioms and rules of deductive inference. This is called undecidability. There are already a host of mathematical propositions whose undecidability was established during the last decade (6). Classical mathematics had its share of impossibility results too; we cannot trisect an angle by Euclidean means, we cannot find a formula to solve exactly all polynomial equations of degree greater than five, we cannot square a circle. The new results in basic mathematics are analogous to these old ones but far more general. Instead of saying that some individual problem cannot be solved, they are saying that whole classes of problems cannot be solved. They are, in a very fundamental sense, a statement of certain limits on man's intellectual ability within the framework of purely axiomatic, deductive systems. Each undecidability proof requires construction of a model in which the proposition in question is true and of another one in which it is false: the undecidability of the proposition follows from the existence of such models, for no general proof or refutation will be possible if the proposition is, in fact, true in some models while false in others. If the logical problem can be related to scientific propositions that are based on empirical knowledge, a decision between the models can be envisaged. The natural scientist would prefer the model that is consistent with nature, and thus the proposition in question is "true" if it is true in the model that is consistent with nature.

In this sense a decision concerning proof or refutation of a mathematical proposition can come from empirical knowledge. This is another example of the empirical solution of problems that cannot be solved by pure thinking.

We conclude that the axiomatic, deductive system of formal logic (including mathematics) is an indispensable aspect of any scientific language. The justification for this statement is neither evident nor trivial. The universal applicability of logic to the description of nature cannot be proved by purely logical arguments. It can only be understood if we refer to the theory of evolution (for a more detailed treatment of this problem see the Epilogue, Epistemology and Evolution).

We briefly return to the multiplicity of natural languages as compared to the universal symbolic languages of the sciences. (Remember that the scientific community is truly international.) To what extent is thinking dependent on the nature (vocabulary and grammar) of a natural language? The natural languages probably exert a considerable influence on the way of thinking of the people. This is one reason why the multiplicity of natural languages is not a satisfactory basis for an enterprise such as science, which must be truly universal, world-wide, and may not depend on the particularities of a natural language, or political situation, or cultural and economic environment. We have developed the view that there is something that can be obeyed by every intellectually capable person: the rules of formal logic, including the axioms, and theorems of the axiomatic-deductive system we call mathematics. We know, furthermore, that the modern sciences and mathematics have arisen within the framework of the Indo-European dialects. From this core of natural languages the symbolic languages of the modern sciences have originated (7). Some people feel that this is one reason for the world-wide unanimity in the scientific description of the world. If a modern Chinese or Turkish scientist describes the world in precisely the same manner as a British or German scientist, this means that they have taken over completely and in toto the Western style of rationalization of the world; it does not imply that their natural languages would have led them to the same system of symbolic languages and consecutive rationalization of the world (5). As an example, the inflexible written Chinese language in which a new symbol is necessary for each new concept has clearly not been useful for the development and communication of scientific propositions. The Latin alphabet, the Arabic numbers, the English language, and in particular the symbolic mathematical language proved to be far superior.

While the excellent linguist WHORF (7) argues that the structure of language plays a role in determining a "world-view," NOAM CHOMSKY, a renowned present-day linguist, feels "that studies of linguistic relativity are entirely premature" . . . and he adds "that one must take a rather skeptical attitude towards the available data regarding language and cognition and their interrelations, whether it comes from the study of normal adults, from pathology, or from developmental studies" (8). Clearly, the questions about "language and cognition" are classical ones [see (9,10)]; however, the issues are still obscure and the experimental studies inconclusive (8).

Science is interested in clear and precise statements. The precision of any statement in any language (including symbolic languages) is limited by the precision of the terms used in the statement. Thus the use of precisely defined terms is essential for science. This does not imply that definitions must be rigid and forever. It implies, however, that definitions are clear and distinct at any given time and may not easily be confused or mistaken. There are a large number of definitions that are in a strict sense essential for science and that should be adhered to vigorously, such as length, mass, electrical field strength, etc. The recently developed International Standard of Units System (SI), which has been rapidly adopted by all scientifically advanced nations, shows the eminent importance of standardized definitions for maintenance and development of science and technology. The use of vaguely or sloppily defined terms is the cause of many misunderstandings, errors, and serious confusions even in science. The biological sciences are particularily endangered at present since a great number of people from the physical sciences and some from psychology have entered the supposedly green pastures of biology. Most of them are not familiar with biological terminology, in particular regarding comparative biology and developmental biology. The "molecular collapse" of some disciplines can only be avoided if the traditional, biological terminology is strictly maintained and sharpened according to the complexity and specificity of living systems. The terminology developed to handle adaptive enzyme formation in *Escherichia coli* is not sufficient to handle differentiation and behavior in multicellular systems. It is part of the responsibility of the scientist to emphasize the importance of a correct and satisfactory terminology. "Satisfactory" means, satisfying the complexity of the system. One can hardly exaggerate this responsibility since without the basis of sound terminology any further intellectual or material investment in a particular field is idle. Goodwill is not sufficient for a person concerned with defining terms. To define terms is an intellectual technique that can be learned and must be exercised (11).

Terminology differs greatly from discipline to discipline depending on the particular part of the real world with which a particular discipline is concerned. Since logic is the same everywhere in science, the diversity of scientific disciplines manifests itself predominantly in the diversity of terminology. We will return to this problem when we consider the possibility or impossibility of reduction between or within the different sciences. At present we only emphasize that the logicians can only give a formal prescription (instruction) for defining terms. As far as the contents of the terms are concerned, the particular expert bears the full responsibility for a satisfactory and useful elaboration of the terms. In particular, a useful term in science should be invariant against the context (whereas in politics, in the ideological confrontation as well as in fields such as sociology, the context variance of a term, e.g., democracy, is skillfully used by the orator or demagogue).

In formal logic a predicator is a word assigned to a thing. A term is an explicitly introduced predicator. Scientific terms are normed predicators (in order to increase precision and to reduce ambiguity). Terminology is a system of terms with rules for the relationships between the individual terms. The process of norming a predicator is called defining.

Definition for Aristotle is a term signifying the set of fundamental attributes that together make up the essence of the thing defined. The definition of a circle might be "closed plane figure with every point equidistant from the center." All circles, and only circles, admit of this predicator (and conversely); moreover, all characteristic properties of circles presumably follow necessarily from it. According to Aristotle every strict definition could be divided into two distinct parts: the one dealing with the genus and the other showing the specific difference by which the given subject varies from others of the same order. Whereas Kant and the modern logicians gave up the strict distinction between genus and difference, the Aristotelian prescription is still, in essence, the core of the logics in taxonomy.

As far as modern sciences are concerned, we are concerned mainly with three categories of definitions: nominal definitions, real definitions, and operational definitions.

In a nominal definition, we replace a laborious expression (the definiens) by a brief expression, in general by a single term (definiendum), without change of meaning. (A "car" is ——; a "gene" is ——). A nominal definition for "science" was given previously (p. 30).

It is obvious that terminology cannot be based exclusively on nominal definitions. Nominal definitions become possible only if a certain number of terms are already available. For the foundation of any terminology we depend on real definitions in science as well as in any natural language (I point at a thing and state "This is a diagram"; "This is an elephant"). Of course, "diagram" and "elephant" can also be defined by a nominal definition if a definiens for these things is available.

A definition on the theoretical level (a "gene" is ——) will necessarily be a nominal definition, whereas a definition on the observational level (a "car" is ——) can be nominal or real. Many terms (and units) in science are operationally defined, that is, by a procedure, by an operation: lengths are defined by rigid rods or (secondarily) by the wavelength of light. Masses are determined by weighing, followed by a theoretical conversion from weight to mass. With this latter example we introduce the general case that an operational definition does not only involve a material (instrumental) operation but also a logical (mental) operation that includes the knowledge about a law of nature (in the present case, proportionality of mass and weight) (5). Operational definitions (and the resulting operational criteria) are of basic importance in science.

As an example, the photomorphogenetic photoreceptor phytochrome was originally defined operationally (Table 1). If an "induction" of a photoresponse by a brief red light pulse is fully reversed by following it immediately with a corresponding pulse of far-red light, the light effect is due to phytochrome by definition (17). Of course, phytochrome was always conceived as a theoretical entity (some kind of a photochromic molecule) but until 1959 it could only be defined by the operational criteria in Table 1. The phytochrome example shows that without operational definitions science would degenerate into speculation. However, in the absence of theory-bound (constitutive) definitions science would remain captured within the range of observable facts and would be simply cumulative (5). We must realize, from the beginning, that there is no significant piece in science that is completely free from "theory."

This is not only true for advanced science but also for daily life. The way we gather information depends basically upon our general, unreflected, prescientific view of nature, e.g., we expect consistency in nature, regularity, order, and causality. We assume that the world out there is real and intelligible. The latter implies that there is a possibility to decide between true and false statements.

A traditional difficulty in scientific terminology stems from the circumstance that science includes theory as well as observation. Is the language (terminology) of observation compatible with the language of the theory? How can the language of the observation records be translated into the language of a paper in a technical journal? While the problem of demarcation of theoretical terms and observational terms and the way they are connected with each other by correspondence rules, has been lucidly elaborated by some philosophers of science, e.g., Carnap, scientists are often amazingly careless in this respect and thus contribute to confusion and misunderstanding. In developmental genetics, for example, speaking of traits or phenotypes is speaking about observation. A statement such as "The pea plant is tall" is clearly a statement in observational language. However, quite often such an observational statement is followed by statements such as "tallness is handed down from parent to offspring," which is at best an unfortunate metaphor (5). Tallness can be measured

Table 1. Operational definition (operational criteria) for the involvement of phytochrome in a response [after (17)].

Light Treatment	Degree of Response
Red light pulse	a
Far-red light pulse	b
Red light pulse + far-red light pulse	b

(observed) but it cannot be handed down from one generation to the next. Here the theoretical term "gene" is unavoidable.

I want to treat the problem of observational and theoretical language in a somewhat wider context by considering the notions of construct and entity, which are both basic for any scientific language. The notion of "constructs" as emphasized by MARGENAU will be considered first (12). Valid constructs are those intellectual inventions that are *useful* in organizing the real world. So-called rules of correspondence link observation to the constructs. A tree is a construct. In MARGENAU'S words: "The tree is real because it is the rational terminus of certain rules of correspondence having their origin in sense impressions and because it satisfies the demands of consistency which common sense imposes."

In my opinion, every useful or valid scientific term is some kind of a construct, an intellectual invention that satisfies the demands of our genetically inherited foreknowledge about the world in which we are going to live. There is no reason to identify stimuli one-to-one to sensations. Rather, we always must take into account the high variability among human beings. We do not "see" stimuli. Our knowledge of stimuli is highly abstract and theoretical. The variability in the stimuli-sensation-construct relationship within human populations must not be too large since this would lead to individual and social solipsism. Since it is generally assumed that in the genetic evolution of man a strong selection pressure was at work with regard to cooperation among the individuals of the population ("inclusive fitness"), we may equally assume some selection pressure toward a decrease of genetic variability with regard to the stimuli-sensation-construct relationship.

Valid constructs are entities, and since this latter term has been used extensively in physics I will use "construct" and "entity" more or less interchangeably.

Constructs do not form a closed class; rather there are different types depending on the distance from the observation point. The simplest and least complicated constructional entity is an easily observable external object. External objects, as carriers of observable properties, are examples of a class of constructional entities that MARGENAU designates as systems. While this is a somewhat loose use of the term system, it is a valid observation that, throughout science, systems themselves are not measurable, but the properties that characterize them are. Thus, the tree is not measurable, the height of the tree can be measured. The electron (the construct) is not measurable, but one may measure its momentum; the atom is not measurable but its atomic weight *is* measurable. The chlorophyll molecule is not measurable, but its absorbance is measurable; the gene is not measurable but we may measure its mutability or even its random nucleotide composition. A man (the construct) is not measurable, but we may measure his weight or his intelligence. We will return to the significance of systems and system properties when we discuss the goals of physiology. However, we keep in mind that the intellectual construction of systems that serve as carriers of certain precisely measurable properties is probably the most important intellectual activity in science.

Constructs need empirical confirmation to become *valid* constructs. Since the constructs form organic parts of suggested hypotheses and established theories, the process of validation (or falsification) of the constructs used in the genesis of a hypothesis or a theory coincides with the empirical validation or refutation of the hypothesis or theory.

It is surprising that a construct, such as an atom, can be approximated by simple intellectual models. Indeed, the billiard ball atomic theory works quite well within a certain range of forces (13). Only recently the physicochemical billiard ball model of the molecule (i.e., the molecule as a sharply distinguished unit) has been replaced in some cases by detailed 3-dimensional models based on electron microscopy and, in particular, on X-ray diffraction analysis. It is generally felt that an insight into the fine structure of the construct "molecule" such as provided by electron and X-ray studies will push the construct closer to "reality." It is sometimes stated that "molecules really exist." This kind of statement might be emotionally attractive. The fact is, however, that the reality of a molecule is the reality of a primarily theoretical construct: we cannot grasp it directly, we can only measure some of its properties.

A very influential construct in biologic theories has been the "cell." The models approximating this construct look very different, depending on the interest of the investigator or the teacher. A cell model illustrating the coexistence law of water potential differs in appearance from the cell models used to emphasize the concept of apparent free space or the concept of membrane flow. All of them claim to represent the construct "cell," which we believe to recognize when we look through the microscope.

I have just used terms such as cell and apparent free space. We all feel that these two terms belong to different levels: the cell is an observational construct (at least if a microscope is at our disposal); the apparent free space is a theoretical construct. Therewith, we return to the language problem proper.

Obviously there are theoretical constructs (or entities) such as molecule, apparent free space, electron, gene, genotype and observational constructs (or entities) such as tree, cell, photographic plate, trait, phenotype. The problem seems to be to define the rules of correspondence, the so-called bridge principles that connect the observational and the theoretical levels (14). However, it turns out that there is no *clear* demarcation between the different classes of entities; the classification of a particular entity may change depending on changing boundary conditions. It is stated that a *theoretical* entity is a thing closely tied to a *particular* scientific theory whereas an observational term is not. However, we recall a previous statement that no observation is free from theory about the world, at least not free from our genetically inherited foreknowledge. Why does the healthy human mind regard a tree as an entity? Why do we reconstruct in our mind *particular* observational constructs from the huge amount of sense impressions?

On the other hand, clearly theoretical entities, such as molecule, may approach the status of an *observational* entity, depending on the technologic progress and on the definition of observation. The shadow of DNA molecules becomes observational in the electron microscope, and in the case of some rDNA it has been claimed that even the *functional* state can be pictured.

Despite these difficulties of demarcation, however, there can be no doubt that these two classes of entities exist and that it is a major problem in science to define bridge principles in order to relate unambiguously the observational language and the theoretical language in a particular discipline. A particularly intriguing case is the dichotomy

gene *phenotypic character*
genotype *phenotype*

in genetics.

The functional gene is largely a theoretical entity; the different characters of the phenotype are largely observational entities even though sophisticated technology (instrumentation) and indirect means of observation (e.g., in viral genetics) introduce a large amount of theory into the observational level. On the other hand, for an entity to be theoretical does not imply that there is always some doubt about whether it exists in reality. In the case of the gene, probably no biologist doubts that these entities really exist, despite the fact that a gene is a theoretical entity, and thus part of a theory.

Every practicing scientist has become used to this permanent problem. We measure absorbance in the spectrophotometer and think in terms of π-electrons (or conjugated double bonds); the next day we again measure absorbance in the same spectrophotometer, but this time we perform an optical enzyme test and thus think in terms of enzyme activity.

As already indicated, the bridge principles between the observational and the theoretical level pose the real problem. For example, in population genetics bridge principles between the genotypic level and the level of the observable characters are rarely stated explicitly (14). This may have the consequence that many investigations in population genetics are empty mathematical games on the theoretical level, no longer related to the observational level, or it may nourish the suspicion that the bridge principles are sometimes adjusted post factum in order to justify the theory in view of observations. In order to maintain the credibility and significance of studies in population genetics, it is necessary to state the assumed bridge principles between the phenotypic level and the level of the genes explicitly and in advance in order to avoid the suspicion of a post factum connection.

In view of the strong, specific effect of some environmental factors on gene expression, any statements about the bridge principles are anything but trivial. If population genetics is applied to the theory of evolution, the bridge principles gain relevance. We must keep in mind that selection is based on phenotypic characters. Selection is an event at the level of the phenotype. In order to treat selection on the gene level, "bridge principles" must be defined and defended explicitly. Otherwise, a theoretical claim with regard to changes on the gene level (e.g., genetic adaptation) cannot be regarded as a fully valid factual claim about the real world. If, e.g., selectively neutral mutations are possible (which is a matter of dispute in present-day genetics), then the principle of selection does not apply at the level of DNA or proteins in the way it does at the phenotypic level (15).

The high prestige of population genetics must not be judged from the massive mathematics applied on the theoretical level; the value of these mathematical models depends largely upon the precise and convincing definition of those principles that bridge the gap between the observational and the theoretical levels. Since population genetics is becoming more and more relevant even for practical purposes, this is a very important task not only from the academic point of view.

It is assumed, usually as a matter of course, that in *Mendelian* genetics the bridge principles would be obvious and without doubt. This, however, is not the case. A Mendelian gene, e.g., is not simply related to a particular character but to the *difference* between the expressions of a particular character. Even in those cases in which the loss of a gene leads to the loss of a character, this is the kind of relationship involved. We will return to this problem in Lecture 6, on the principle of causality. At the moment, I only want to emphasize that we eventually must eliminate those misleading or at least imprecise statements such as "it is sufficient to define the gene as a unit transmitted from parent to offspring which is responsible for the development of certain characters" from leading textbooks (16).

I would like to close this lecture with a brief look at the progress of science. The progress of science (as judged by the increasing reliability of theories) is closely related to the use of "better" terms and constructs. The progress of science has always been connected to the progress of terminology. This is not a logical problem but concerns the *contents* of the terms and constructs. Intuition and creativity of the scientist play the major part here. The introduction of precisely defined terms (and therewith concepts) such as mass, force, momentum, and energy have had a tremendous influence on the progress of physics after they had replaced the old term vis. The term entropy was indispensable for the rise of thermodynamics, the construct gene has revolutionized biology as has the construct quantum the science of physics. We realize that the terminology and the system of constructs must remain flexible. Terminology in vivid science is an *open* system and thus able to progress. However, as every open system, it is also menaced by decay if not maintained properly. Maintenance includes not only the continuous input of intellectual effort and creativity to improve the system but also the defense against carelessness and slovenliness and the defense against those irrational forces that try to weaken the *strength* of scientific language and therewith, scientific thought.

On the other hand, there is a risk that constructs become doctrines and written knowledge becomes idolized. Scientific thought is not free from this all-human tendency.

4th Lecture

The Scientific Approach (2): Data, Hypotheses, and Theories

Science is an organized venture of the human mind that aims at genuine knowledge. The basis of scientific knowledge is true singular propositions (judgements, statements, data, facts),[1] which are based on observation (or, in more general terms, justified by empiricism). A statement is true if it refers to an existing state of affairs in the real world; a statement is false if the state of affairs to which it refers does not exist. Every unambiguously formulated statement is either true or false: there is no third possibility. But even a false statement may contain some truth. If it is 12^{30} and I am told it is 12^{25}, this statement is false, but less false than the statement, it is 12^{15}. It is essential that any statement (judgement) is strictly intersubjective, i.e., it must be formulated and communicated in a way that it can be scrutinized (verified or refuted) at least potentially by every competent member of the scientific community. The consensus principle is essential for science (1).

We recall the previous definition that science is the study of those judgements concerning which universal agreement among competent scientists can be obtained. It is always (at least implicitly) assumed that these judgements are *logically* correct. This implies that only those conclusions and steps of thinking have been used in the elaboration and formulation of the judgements that are permitted by formal logic. Thus, logic always plays an indispensable role in the creation of basic statements insofar as it is the tool to control the correctness of the steps of thinking involved. Scientific judgements are subject to logical scrutiny. Thus, it is logic that separates the controlled thinking of science from blind phantasy, irrational imagination, and willingly illogical statements. With regard to the *contents*, the

[1]Statements pertaining to facts may be called singular propositions, factual propositions, statements of facts, basic statements, basic judgements. In scientific practice these statements are often called "data" or simply "facts."

truth of a logically correct statement is a matter of empirical scrutiny. Logic cannot judge whether or not a judgement is in accordance with nature.

Basic judgements (data, facts) are never free from *theoretical* implications. Nobody in nontrivial science simply gathers facts without being influenced and even guided by a conceptual framework. Thus, even the basic judgements already reflect some theoretical commitment of the scientist or at least some particular way to look at nature. Mendel is an outstanding example. When he designed the experiments described in his 1865 paper he must have already arrived at his theory. Otherwise it is difficult to understand why he decided to do those particular experiments. In conclusion, at least in advanced science nobody starts from scratch. All obtaining and processing of data, every use of instrumentation, is embedded in a theoretical framework whether we realize this explicitly or not.

The operational definition of science as the study of those judgements concerning which universal agreement (among competent scientists) can be obtained implies that scientifically relevant knowledge is public knowledge in the sense that our subjective knowledge must be written down, communicated, and criticized by our fellow experts before it becomes scientific knowledge in a strict sense. The reasons for "public" knowledge were particularly well stated in a report of the AAAS Committee on Science in the Promotion of Human Welfare (2): "Free dissemination of information and open discussion is an essential part of the scientific process. Each separate study of nature yields an approximate result and inevitably contains some errors and omissions. Science gets at the truth by a continuous process of self-examination which remedies omissions and corrects errors. This process requires free disclosure of results, general dissemination of findings, interpretations, conclusions, and widespread verification and criticism of results and conclusions." As a consequence of this notion of scientific knowledge as public knowledge, there are pioneer areas in science, characterized by relatively private or personal knowledge regarding data or conjectures, which may not yet be considered as parts of "solid science." An analogy may illustrate this important point: during the invasion of North America by the white settlers there was a frontier range with more or less independent settlers, which no longer belonged to the completely unknown (the Indian country) and not yet to the organized state. This is analogous to the pioneer range of science, which is occupied by bold conquistadores, roaming trappers, and half-civilized settlers whose activities are absolutely essential for the scientific progress but whose statements require consolidation, i.e., the consensus of the scientific community to reach the status of truly *scientific* statements.

In a scientific field, predominantly ruled by paradigms, such as classical thermodynamics, atomic spectra, or comparative anatomy, the pioneer range may only be small; in revolutionary times, however, the pioneer range becomes very wide and of crucial importance—and can even endanger the organized state—a matter that we will take up again in Lecture 10 on Tradition and Progress in Science.

At this point of the discussion I would like to make a firm statement with regard to the responsibility of the scientist: A *primary* responsibility of the scientist is to contribute to the wealth and reliability of data (= basic judgements, singular propositions). Those basic judgements that have been subject (at least potentially) to the scrutiny of the scientific community and found to be true are called facts or objective data. Fruitful hypotheses can only be based on objective data.

This is intuitively felt by every good scientist. It does not matter that most scientists do not pay attention to explicitly formulated methodologic principles. The important point, however, is that they *behave for some reason or other in accordance with these principles*.

The scientific literature is replete with instances of misstatements, and oversimplifications in secondary sources become accepted as "facts." This is particularly true in fields in which public interest is high, e.g., in cancer research. There is no simple cure! Every author of a review article or a textbook knows how difficult it is to simplify without simplification and to integrate information from a large number of original articles without any severe misstatement.

Three years ago, in an editorial in *Science*, J. ROSS MACDONALD (3) answered the question "Are the data worth owning?" with an embarrassing and costly "No" for a major fraction of the published scientific and technical data. Since then—as COMPTON has recently assured us (4)—"slow but steady progress has been made in increasing the reliability of data which is so essential for the orderly conduct of research and development programs." It is not only the total research and development effort (see Fig. 22) for which reliable data are indispensable; rather it is the scientific enterprise as a whole that depends on high quality data. Objective data of high quality and reliability are the pillars on which science is built.

The next step we have to deal with is difficult to describe. This stage is conventionally referred to as inductive inference or simply "induction" (Fig. 1). It is the step from singular propositions to general propositions, or, the step from basic judgements or data to a hypothesis (and eventually to a theory). The difficulty is that most experts feel that this step cannot be completely rationalized. The formation of a fruitful hypothesis is not only a creative act but also something like a mystery that includes factors such as "intuition," "foreknowledge," "constructive power," "inspiration," and "having an idea." A tentative *description* of the process can be given as follows: We have the desire to "understand" (or to "explain") some data (basic judgements), e.g., we have the desire to understand why an eclipse occurs or why we get an upset stomach after we take aspirin. In science we assume that the basic judgements we have in mind represent objective knowledge about a real system that exists "out there" in the real world. An explanation of the singular propositions (basic judgement) means that

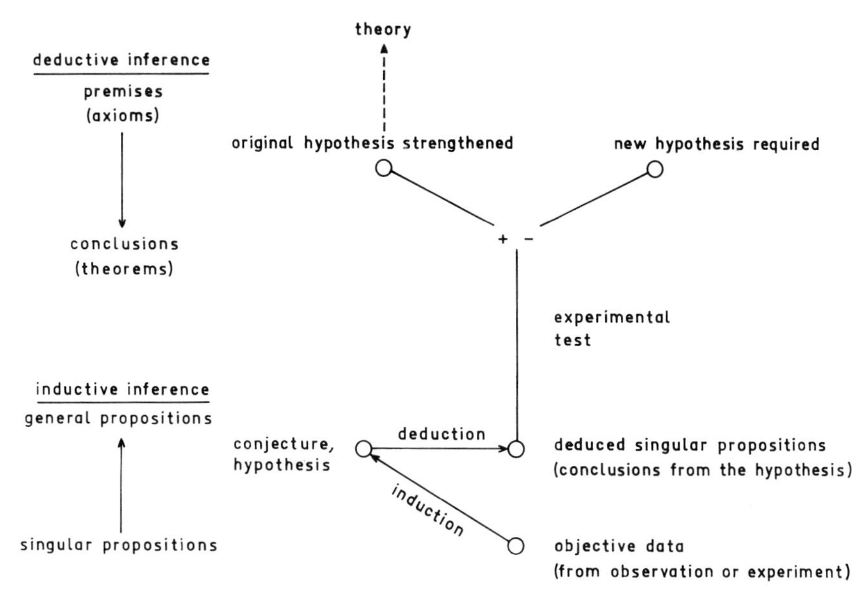

Figure 1. A sketch to illustrate interplay between inductive and deductive inference and experiment in process of scientific discovery [after (12)]

these singular propositions can be shown to be logical consequences of the natural system we have in mind. Since the real system is not directly accessible we construct a *model* of the real system using the information of the relevant singular propositions (facts, objective data) in addition to our more general experiences we have made with nature and to our intuitive "knowledge" about nature. A tentative model of this kind is called a hypothesis or a "conjecture" [to introduce POPPER'S term (5)]. Since a hypothesis is constructed with the aim of understanding nature, it is emphasized that the hypothesis *represents* at least some aspects of a real system and thus *refers* to a more or less large section of the real world. From the informational point of view a hypothesis is a *coherent* system of more or less general propositions. A hypothesis is thus an intellectual construction, developed to explain facts (singular propositions) and intended to represent in some preliminary manner a certain section of reality. These are the indispensable properties of a useful hypothesis: It must be consistent with logic, e.g., it must be free from obvious internal contradictions, and it must offer a satisfactory "explanation" for those singular propositions whose information content (objective knowledge) had been the basis and the *motivation* for the construction of the hypothesis. Remember that we were motivated to make a hypothesis because we had the strong desire to explain something.

What does explanation mean in this context? It means that we deduce by means of deductive inference conclusions from the hypothesis that are identical with the original singular propositions. Since all the particular conclusions were already implicit in the general propositions of the hypothesis we need only to draw them out in explicit form. This is done with the same deductive logic that is used, e.g., in mathematics, to draw conclusions (theorems) from the premises (axioms).

I do not intend to discuss the structure of purely deductive systems any further since this is part of the theory of logic with which you became acquainted, at least implicitly, when you received your introduction to mathematics.[2] Rather, I want to apply the structure of an axiomatic-deductive system to the function of a hypothesis. We regard the propositions of the hypothesis that have the character of general propositions (something like tentative laws) as being analogous to logical premises or axioms. The statement: "The hypothesis explains the facts (singular propositions, basic judgements, objective data)" means that the singular propositions can be deduced as theorems from the general propositions of the hypothesis, which function as axioms (or premises). The really decisive point is now the following: From the general propositions of a hypothesis (always considered as analogous to axioms), we can deduce more singular propositions (theorems) than we originally had available. These purely deduced singular propositions—anticipated facts—can now be compared with nature, as a rule in an experiment. If the experimental (or observational) check leads to a positive result (i.e., the derived singular proposition is indeed a true statement, referring to an existing state of affairs in the real world), the hypothesis is strengthened (supported, confirmed). If, however, the experimental check leads to a negative result (i.e., the derived singular proposition is a false statement since the state of affairs to which it refers does not exist), the general propositions of the hypothesis (or at least some of them) cannot be true. As a consequence, the hypothesis must be modified or even completely declined. Any intelligent hypothesis, even though it was subsequently refuted, is never quite useless. It has helped to deduce new singular propositions and thus to predict events so far not measured, thereby stimulating experimentation and observation. In any case, to shape a *valid* hypothesis any contradictions between the general

[2]A fine example of deductive inference is the syllogism. A syllogism consists of three propositions, two forming the premises and the other the conclusion. Each proposition can be expressed by a subject and a predicate connected by the verb to be, the copula, and if we call everything that stands as either subject or predicate a term, there must be three and only three terms in the syllogism. The one common to the two premises is called the middle term, and it is on this common element that the inference depends. The other two, having been connected by means of this term, occur without it in the conclusion. Thus in the usual example of the syllogism,

All men are mortal	
Socrates is a man	premises
Socrates is mortal	conclusion

man is the middle term connecting Socrates with mortality so that we could, even if he had not already died, know that he was mortal (13).

propositions and experimental data must be eliminated. A hypothesis that has survived many scrutinies of this kind is called a theory, at least if the general propositions refer to a considerable segment of the universe. Thus, a theory is a systematically organized system of general propositions that permit the explanation (in the foregoing sense) of many singular propositions and thus "describes" a considerable segment of the real world. Admittedly, this has been a very *formal* description of those acts that are involved in the formation of a theory. In practice, e.g., in the practice of a creative laboratory, a precise distinction between inductive inference ("guess work," "having an idea") and deductive reasoning (drawing testable conclusions from the hypothesis) can hardly be made, since both processes may occur almost simultaneously and with continuous feedback, e.g., in a discussion among experts about the fruitfulness of a hypothesis. The careful experimental checking of a hypothesis requires material preparations. This act is, therefore, usually separate in time from the intellectual development of the "conjecture."

Even "having an idea" or "shaping a hypothesis" need not be the work of a single mind. It is often difficult to decide in retrospect who did have "the idea." In any case the experimental *testing* of an idea (the "check work") is usually a process to which several or even many skilled hands and brains may contribute (6).

The value of a theory can be judged by its predictive power, i.e., a "good" theory is a theory that predicts a large number of singular propositions that are accessible to an empirical check by observation and experiment. The expressions "predictive power of a theory" and "explanatory power of a theory" are largely synonymous. An obvious difficulty in deriving singular propositions that are supposed to be checked empirically from general propositions lies in the more or less different terminologies used on the theoretical and on the observational levels. The singular propositions (in order to be scrutinized empirically) must be formulated in the language of the workbench without any distortion of meaning or loss of unambiguousness. This requires a high degree of intellectual skill and discipline (see previous lecture).

A hypothesis is called sterile if it does not permit predictions on the observational level and appropriate experimental tests; a hypothesis is called fruitful if it *does* permit predictions that can be checked on the observational level. It is obvious that science is only interested in fruitful hypotheses. Many speculations in philosophy and theology bear the stigma of sterile hypotheses. They may sometimes be sources of consolation or strength for man, but from the point of view of epistemology they may be ignored.

In biology, the terms "theory" and "speculation" have often been confused, in particular by those experimental biologists who tend to discourage theoreticians in favor of what they call "solid science," largely a compilation of theoretically unrelated facts. Actually, also in experimental biology progress depends on precise measurements *and* on internally consistent conjectures. Speculations are "theoretical" statements that cannot be subject to experience; as long as a theoretical statement is internally consistent and can potentially be tested on the observational level it is legitimate, irrespective of its boldness. The decisive point is that correspondence to the facts can be proved. Any statement belongs to science if it is in principle refutable. Bold and imaginative conjectures are a major drive of scientific progress, but they must be controlled by severe criticism including severe experimental tests.

Theories may be grouped as good and poor ones. As indicated previously, a good theory covers a wide range of phenomena, i.e., takes care of many basic judgements (=data) and must be testable in clear-cut experiments, i.e., the prediction must be precise and unambiguous and the actual testing must be feasible with the presently available instrumentation. Sometimes, the very strong desire to test a theory may even lead to the development of more sophisticated instruments. The more precise the prediction, the higher the gain of prestige for a theory if it is confirmed experimentally. A characteristic of poor theories, found predominantly in the so-called social and political sciences, is that their predictions are so vague and diffuse that clear-cut refutations become impossible (6). Astrologic predictions are of this kind, as are predictions from the so-called historical materialism (or, scientific socialism). While the earlier Marxist theory of history was quite specific with regard to predictions (e.g., of the coming social revolutions) and was promptly refuted by the historical process, the modern Marxist "theory" is not only clumsy (i.e., loaded with auxiliary hypotheses) but so diffuse that its predictive power is virtually nil. Such theories are uninteresting for the scientist.

Any general proposition within the framework of a theory may be called a law. As described previously, all scientific laws are empirical insofar as they permit the deduction of singular propositions that are empirically scrutinized. Sometimes the expression "empirical law" is used in particular for such general propositions, which can be tested by direct experiment. However, even an "empirical law" of this kind is always embedded in a theoretical structure and thus is not principally different from a more "theoretical law."

Biological theories, laws, and predictions are not fundamentally different in their logical structure from those in the physical sciences. This aspect will be dealt with in more detail in a forthcoming lecture, in which we will investigate the problem of reductionism in the relationship between biology and physics. At this point, however, it must be admitted that the kinds of predictions in physics and in biology are sometimes quite different. Two examples will suffice to illustrate this point.

Knowing the laws that govern the solar system and defining the starting condition (initial conditions), one can predict the future position of a planet with extremely high quantitative precision. This is the clasical case of prediction in science. The theory implied is relatively simple and very precise.

In biology it often sounds like this: we know that the chromosome theory of inheritance is generally valid among eukaryotes. Man is an eukaryote. Therefore we can predict that a deletion will lead to serious physiologic and psychological deficiencies in man. This kind of prediction sounds vague. However, the prediction is precise insofar as it does not permit exceptions, and the relevance for man is very strong.

Sometimes philosophers demand that every *scientific* theory that really deserves this title must be a *unified* theory (such as Newtonian mechanics), but this claim is not appropriate. At least some theories in vivid fields of science are still developing and may consist of subtheories that are not fully compatible or even rival. It would be far too restrictive to narrow down the term "theory" to a mature and nearly closed axiomatic-deductive (physical) theory and thus exclude nearly all biological "theories" from the dignity of this term. Anyhow, in principle all scientific theories are subject to progress. A theory is correct ("true") as long as there is not objective data that contradicts the theory. Therefore theories in science are in principle open, progressive intellectual constructions. A "true" theory (as the term is commonly used in science) is one that shows a high degree of inner perfection and has never been refuted despite many sharp experimental tests. In more subjective terms: we believe strongly that the prediction from a "true" theory will always be confirmed by subsequent experience. However, in principle, every theory is open to future refutation, and it can be superseded by a better one. It is never true in a *metaphysical* or ontologic sense. The notion of truth is clearly operational in science, both on the level of singular and general propositions. POPPER (5,7) has emphasized the point that theories cannot be verified definitely; they can only be made stronger by failures to refute them. But in principle every theory remains fallible, and attempts to disprove theories are a major concern of science—at least in the eyes of this leading philosopher of science.

As practicing scientists, we enjoy much more "verification," at least if it concerns a hypothesis or a theory to which we have contributed. (It is a different matter if we investigate a hypothesis advanced by our scientific rivals; in this case a successful refutation can be a very positive emotional experience.) Irrespective of emotions *pro and contra*, we are all aware that the refutation of established hypotheses and theories is related to scientific *progress*. It is indeed our profession to eliminate errors. However, for the experimental scientist refutation is not a simple thing. Apparently refuting data may result from a false statement or from false assumptions that were introduced during the performance of the test. It requires strong, unambiguous experimental data (objective data in a very strict sense) to refute a theory that has been satisfactory for a long period.

Errors and even deceptions are by no means uncommon in science. They happen in the elaboration of data not only as (inevitable) random errors but also as (principally avoidable) systematic errors. The most serious errors are those that occur in the *logical* operations involved in testing a hypothesis. Often the experimental part of a test procedure depends crucially on the state of the techniques involved. This becomes particularly important if sophisticated multistep techniques are required to perform a test experiment. These are a few reasons why an established theory should not be rejected simply because some data do not fit. It must always be presumed that the data in question are not truly objective.

If there is a lack of agreement between the theoretical prediction from a theory and the experimental data, the experienced scientist will first check the logical operations involved in the prediction and then the performance and the boundary conditions of the experimental test. Only if he is convinced that no error is involved and that there is no other explanation for the discrepancy between established theory and experimental data will he (or rather, the scientific community, since the consensus principle is strongly implied here) consider the possibility that the theory cannot be maintained. Premature "refutations" of theories have occasionally created much turmoil in science. It is clearly part of the responsibility of the scientist not to proceed too quickly if the persistence of an established theory is in danger. An example [adopted from DAVIS (6)]: At the end of the 19th century the atomic theory was generally adopted after a fierce argument throughout the 19th century. According to this theory the element nitrogen should be the same whether obtained as a residual gas from the air (after removal of the oxygen and carbon dioxide) or by the decomposition of nitrates. However, consistently in experimental results, nitrogen from the air appeared to be about 0.5 per cent denser. This discrepancy between theory and experiment led Ramsey to propose that the nitrogen obtained from the air contained traces of some inert heavier gas. He did not seriously consider the alternative—that the atomic theory was in danger. You all know that Ramsey's explanation of the discrepancy between theory and experiment was right. He isolated argon, and this opened the way to the discovery of the whole family of rare gases.

Most scientists are quite reluctant to claim a refutation of an established theory. They tend to accept slight discrepancies (with some observation) in a theory if the theory has already gone through many tests. Eventually, however, the theory "will be adapted to the facts" by means of additional assumptions, called auxiliary hypotheses. In general this "adaptation" not only decreases the simplicity of the theory but also its trustworthiness (credibility). Thus the "saving" of a theory by additional (sometimes obviously artificial) assumptions is usually the first step in abandoning the theory completely. In any case, the "patching" of a theory is a signal to look for a better one (6).

Sometimes, however, auxiliary hypotheses have turned out to be very successful. As an example, biologists have responded to the experimentally found exceptions from Mendel's second law (it was found that some genes do not segregate independently) by the auxiliary hypothesis of coupling groups. This auxiliary hypothesis led to the chromosomal theory of inheritance since it was found that the number of coupling groups in a particular organism equals the number of chromosomes in the haploid state (n).

Debates on theory-choice in science can be very hard, and even irrational elements and personal prejudices play a part. I do not know of any discussion about the fruitfulness of competing theories in which the scientists involved would have agreed about the relative fruitfulness of the competing concepts.

As SIMON (8) pointed out, "a scientist's expectations and interpretations of new experiences are a function of his primary commitments, as are his responses to additional information and assertions. Science is methodical, of course, but no application of formal rules of proceeding and no appeal to objective evidence are sufficient to account either for the scientist's commitments or for the behavior that results from them."

This is by no means unscientific. There is indeed no neutral instance, no infallible arbiter, for theory-choice in the sciences that could force any person to accept the choice—as is the case in an axiomatic-deductive system once the axioms and the rules of deductive inference have been established. It is the scientific community of experts that eventually gives preference to a particular theory. But this can be a long and tedious process. Blind stubbornness (of "popes," in particular) can account for long delays in acceptance, but if the number of positive arguments (many instances in favor of the original conjecture, no refutation) has reached a certain threshold level, the validity of a particular theory can no longer be challenged even by a "pope," who has been in favor of an alternative concept.

DAVIS (6) has tried to formalize the process that leads to increasing the confidence of the scientific community in the validity of a theory. According to DAVIES, we can write for our confidence in a given theory

$$\text{confidence odds on} = \frac{f_1(t)}{f_2(g) \cdot f_3(s) \cdot f_4(p) \cdot f_5(r)}$$

where t = testedness; r = refutedness; s = simplicity; p = power to make precise predictions; g = generality. It looks strange at first sight that simplicity and precision are found in the denominator. The reason is (according to DAVIES) that a high simplicity is suspected because we think it may be an oversimplification and that the precision of the prediction will reduce our confidence in it. Generality is found in the denominator because we become less confident in a theory if it tries to cover an enormous range of ideas and concepts.

Most scientists still feel that this kind of consideration is somewhat childish and at best superfluous. However, these critics are ill-advised. It is our knowledge about the structure and confidence of theories that is essential for any further progress in the sciences, in particular in the biological sciences. The low "logical" standard of biology (as compared to physics) is mainly due to the fact that biological literature is overwhelmed by mere reportage of observation, published just as observations without being related to explicitly and precisely formulated hypotheses or theories. Ironically this simple reportage of unrelated experimental observations is sometimes called "solid science," even by leading biologists. In fact, such observations are often scientifically meaningless. They disappear rapidly, mostly forever, in some scientific journal. In my opinion this is not only a waste of public funds and of facilities for scientific publication (which the taxpayer will not tolerate forever), it is also a serious indication of the lack of maturity in most biological fields. It has the further consequence that people are constantly "discovering" what is already "known" because the original discovery has been forgotten (or is simply not known to the investigator *and* to the responsible editor and reviewers).

As already pointed out, the formulation of hypotheses should not be confused with sheer and fruitless speculation. The value and status of a scientific hypothesis must be measured by its predictive power, i.e., to what extent the hypothesis permits precise predictions that can be subject to rigorous experimental tests. Also in biology, we should restrict (as a rule) the term discovery to the provision of an adequate hypothesis to cover observations so far not "understood," i.e., not explainable by a theoretical framework.

I would like to return to a more *positive* aspect with the statement that the history of successful theories documents very convincingly the observation that objective data are indeed the ultimate court of appeal in science, irrespective of prestige, illustrious names, and sacred tradition. This is a tremendous achievement in view of the power of authority in the Middle Ages *and* during the ascent of science.

The Newtonian theory of gravitational forces is still the masterpiece of a scientific theory—elegant, comprehensive, simple, and at the same time, relatively accurate—although we know that there are significant deviations from the theory under extreme conditions. Under less extreme conditions the theory works adequately. The prediction of the existence of the planet Neptune was a particularly spectacular triumph of Newtonian theory.

You probably remember that at the beginning of the last century astonomers realized that the orbit of the planet Uranus showed slight discrepancies with the predictions derived from Newton's theory. Adams in Cambridge and Leverrier in Paris explained the observational situation by the auxiliary hypothesis that one additional outer planet exists that had so far remained unobserved. On the basis of this hypothesis Leverrier predicted precisely where the new planet should be in 1846. He communicated his prediction to the astronomer Galle at the Berlin Observatory, and on September 23, 1846, Galle found Neptune within less than 1° of arc at the predicted position.

On the other hand, Newton's corpuscular theory of refraction of light from 1704, which was preferred throughout the 18th century to the rival wave theory advanced by Huyghens, was finally refuted in 1850 by Foucault's direct measurements of the velocity of light, which showed that light travels more slowly in a dense medium (6). This was in contrast to the prediction of Newton's theory that it was accelerated. Since Foucault's data could easily be confirmed ("objective data") Newton's theory is considered as being definitely refuted.

The germ theory of infectious diseases as developed by Pasteur states that infectious diseases are spread by microorganisms (and virus, as we may add nowadays). Many apparent refutations of the germ theory turned out to be experimentally wrong, mainly due to incomplete sterilization. While Pasteur's germ theory received countless confirmations (and no standing refutation so far) and has opened a broad avenue for preventive and curative medicine, Pasteur's "theory" about alcoholic fermentation proved to be totally wrong. Pasteur favored the idea that a kind of élan vital, an attribute of the *living* cell, would be indispensable to enable the yeast cell to perform the process of alcoholic fermentation. Buchner (in 1897) finally refuted this vitalistic theory. He showed that a cell-free extract obtained from the yeast cells could perform the transformation from sugar to ethanol. Buchner called the active principle "zymase." This was the starting point for what we nowadays call biochemistry. It was a shift in paradigms; a scientific revolution.

As will be discussed later (see lecture on Tradition and Progress in Science), most scientific activity is concerned with "puzzle-solving" [to introduce one of KUHN'S (9) famous terms] *within* the framework of existing theories rather than with trials to refute and overthrow existing theories. An organic chemist who tries to elucidate the structure of an unknown alkaloid will not doubt the validity of the existing theories in the relevant fields of science. A biologist who wants to use the electron microscope will usually not doubt the validity of those theories on which this marvelous instrument is based. Rather he will take the validity of some "paradigms" (to introduce another of KUHN'S famous terms) for granted and base his "puzzle-solving" on this assumption. The steady growth of science is not primarily a matter of the revolutionary overthrow of leading theories. Rather most of science is modestly concerned with the solving of problems ("puzzles"), including very practical and very urgent problems, within the framework of accepted and established theory. In "normal," i.e., in puzzle-solving science, we follow essentially the same inductive-deductive pattern we have discussed previously as the combination of "guess work" and "check work" in connection with the formation and checking of hypotheses. However the conjectures remain deliberately within the realm of already existing theories. Conjectures *within* the network of established theory represent the decisive characteristic of "normal" science. If a conjecture does not work, i.e., is not in accordance with the outcome of carefully devised tests, we are prepared to discard the conjecture and replace it with another one. The leading theories, the paradigms of our field, will not be touched by this process. Only in rare cases will it happen that new data are definitely not consistent with established paradigms (remember Buchner's zymase) or that a totally new theory turns out to be superior to an established paradigm (e.g., that DNA rather than protein is the material carrier of genetic information from one generation to the next). In these rare cases an exchange of paradigms or of a whole set of paradigms must be envisaged. These rare events in the history of science have been called (primarily by KUHN) "scientific revolutions." We will return to this point in the Lecture on "Tradition and Progress in Science."

Why can the human brain invent valid hypotheses and theories at all? An answer to this old and honorable question of epistemology can be obtained by following the same kind of reasoning we have been using in our attempt to answer the question why logic works in nature. Man is a product of genetic evolution. Without the intellectual ability to create more or less reliable hypotheses, at least about some segments of nature, man would not have survived genetic evolution. Wrong thinking, unreliable myths, irrational superstition, and phantastic metaphysics have been a permanent threat during the ascent of man; on the other hand, right thinking and the creation of fruitful hypotheses about the relevant part of nature must have been a selection advantage of greatest importance during our own genetic evolution. This implies that we have inherited *genetically* (and not only culturally) some "foreknowledge" or instinctive knowledge about how nature works, and we make use of this "foreknowledge" when we create hypotheses.

For this reason the creation of a hypothesis by the healthy human mind is not a random phenomenon. Rather, categorical a priori knowledge will enable us to advance reasonable hypotheses, bypassing the huge multitude of possible hypotheses.

The partial coincidence of the categories of nature and the categories of our mind, the preestablished harmony, has fascinated some philosophers in the past. It is still more fascinating, I feel, that modern science can *explain* why this preestablished harmony must *necessarily* have become established in the course of genetic evolution. As soon as man started to think, he had two alternatives: think predominantly wrongly—and perish, or think rightly (at least to some extent)—and succeed. Only very few human populations, all belonging to the same species, have survived genetic evolution. We may assume that the other branches of the genealogical tree have disappeared mainly for the reason that they could not overcompensate the risk of wrong thinking by the advantage of right thinking and reasoning in accordance with nature.

The modern *Homo sapiens* is by no means safe. We always walk on the verge of extinction. Even today, the tremendous progress we owe to modern science and technology is threatened not only by moral slovenliness and by the decay of values but also, and possibly still more, by the impact upon *rational* thinking of wild irrationalism, uncontrolled phantasy, old-fashioned but fashionable superstition, new mythology and—still worse—with the so-called dialectic thinking—all varieties of that wrong thinking that probably wiped out the other hominids in the course of evolution. It has been extremely risky in evolution to leave the safe shelter of instinct-controlled behavior. It has been a way of no return. The alternatives for *Homo sapiens* sound brutal: think rightly or perish. *Rightly* means, in accordance with the laws of nature and *with* the nature of man. This aspect will be considered further in the Lecture on Epistemology and Evolution.

We have previously mentioned that inductive inference is traditionally regarded as something like a mystery. Is this really the end of the rational discussion about this point? We will consider this problem once more from the point of view of genetic and cultural evolution. With respect to purely deductive systems the logical structure of the relationship between premises (axioms) and conclusions (theorems) seems to be perfectly transparent (the actual movement from the premises to a conclusion can be intuitive, of course; it is the *structure* of deductive inference that seems clear). On the other hand, the structure of inductive inference (the structure of "scientific discovery") that leads to coherent hypotheses is often thought to be inaccessible to logical reconstruction. To quote POPPER (7): " . . . the initial stage, the act of conceiving or inventing a theory, seems to me neither to call for logical analysis nor to be susceptible to it. The question of how it happens that a new idea occurs to a man—whether it is a musical theme, a dramatic conflict, or a scientific theory—may be of great interest to empirical psychology; but it is irrelevant to the logical analysis of scientific knowledge . . . My view may be expressed by saying that every discovery contains 'an irrational element,' or a 'creative intuition,' in Bergson's sense." On the other hand, there exists a fine explication of the intuitive inductive rules used by scientists (10) and there are some attractive descriptions of the *process* of inductive inference, e.g., that elaborated by CAWS (11) with regard to "a great scientist . . . in his creative moments: . . . it involves an ability to keep a lot of variables in mind at once, to be sensitive to feedback from tentative calculations (or experiments), to assess strategies for the development of time and resources, to perceive the relevance of one fact to another, or of a hypothesis to facts." It seems that there is nothing mysterious or irrational about scientific creativity if we look at it in the context of man's genetic and cultural evolution. As CAWS puts it: "Formal logic . . . represents a refinement and specialization of the principles of every day argument; the logic of scientific discovery . . . will similarly prove to be a refinement and specialization of the logic of every day invention." This is indeed the conclusion we reach if we take into account our scientific knowledge about the ascent of man.

There is no *abrupt* breach between the vague prescientific endeavor to know and the intellectually disciplined scientific undertaking to obtain genuine knowledge. Modern science has arisen from primitive natural philosophy and the crafts.

5th Lecture

The Scientific Approach (3): Laws, Prediction, Explanation

The indispensable basis of science is true singular propositions (facts, objective data). However, without the intellectual framework of theories, facts would be purely cumulative and of little meaning. The full power of science becomes obvious only in the construction of theories and in the formulation of *general* propositions or laws. Scientific research is concerned with the improvement of singular as well as of general propositions. We have already described the significance of objective data and the close connection between inductive and deductive steps in the construction and defense of theories (Fig. 1). We now consider the term "law" and its meaning.

Even in the sciences the term "law" designates different things. Sometimes only the general propositions of a theory—which are analogous to the axioms of a purely deductive system—are called "laws"; sometimes, however, even deduced propositions of limited validity—analogous to theorems in a purely deductive system—are called "laws." In other cases, the term "empirical law" designates general propositions that are results of limited inductive inference rather than of deductive reasoning. In any case, the term "law" is supposed to designate some high-level general proposition.

Some general propositions are thought to be valid for the whole of reality. These universal laws are sometimes called *principles*. The first and second laws of thermodynamics may serve as examples for the formulation, interrelationship, and interdependence of principles.

Ideally, laws can be considered to be invariant relations of lesser or wider scope that connect constructs. As an example, force, mass, and energy are constructs; they are connected by the laws $F = ma$ and $E = mc^2$. It is the quantitative, invariant relation between valid constructs that characterizes scientific thought. Low-level general propositions are often referred to as rules. The lowest level of general propositions is represented by empirical generalizations, which are often nothing more than *descriptive* generalizations. These may reflect in many cases accidental coincidences rather than law-like structures. Some premature generalizations in plant sociology and plant ecology are obviously charged with this stigma.

Many descriptive generalizations in biology (sometimes called "qualitative laws") are not, or not yet, integral parts of the network of a theory. It cannot be ignored that many generalizations in biology must be regarded as descriptive generalizations and thus as isolated statements not kept together by a unifying theory. In biology we tend to support empirical generalizations ("high humidity and temperature lead to mould formation"; "flu is favored by wet and cold weather") by some case histories rather than to provide a solid and internally consistent theoretical frame (1).

For the purpose of our present consideration, two laws govern the conditions that must exist if and when energy transformation occurs: (1) The law of conservation of mass states that mass is indestructible. The weight of materials entering into any reaction must be exactly equal to the weight of materials to be accounted for when the reaction has been completed. This formulation per se does not consider the equivalence of mass and energy as expressed by $E = mc^2$; however, it depends on the validity of the principle that mass and weight are proportional to each other (2). The law of conservation of energy, often called the first law of thermodynamics, states that energy is indestructible. Hence the quantity of energy entering into any reaction must be exactly equal to the quantity to be accounted for on completion of the reaction. Again, this formulation per se does not consider the equivalence of mass and energy. If the statement $E = mc^2$ is considered to be a true general proposition of universal validity, the law of conservation of mass is simply another formulation of the law of conservation of energy and vice versa.

The first law of thermodynamics serves as a basis for establishing all energy equations; hence it is truly the basic principle of the science of thermodynamics including bioenergetics. We must recognize, however, that the information from this principle is clearly limited. If heat is converted to work, the first law explicitly evaluates the relationship between work produced and heat consumed, but in itself provides no information as to the possibility of such a conversion taking place. Such information must be obtained elsewhere, usually from the second law of thermodynamics. This law says, in effect, that the availability of energy for work within any system will never spontaneously increase and can never be increased by any external method that does not result in a greater decrease in availability for work of a like quantity of energy.

The first and (in particular) second laws of thermodynamics have been expressed in various forms and in different natural languages during the history of science. It is assumed, of course, that all these statements are *equivalent*, i.e., that they express precisely the same general information, the principle. A modern formulation [adopted from WALKER (2)] shows the tendency for simplicity and unambiguousness. First law: work may be changed to heat at 100% efficiency; second law: heat may be converted to work, but at efficiencies that are always less than 100% and usually less than 50%. The difficulties in formulating these basic principles precisely and in full accordance with the meaning clearly shows the limitation of any *verbal* expression in science. In a purely *symbolic* language an unambiguous and thus satisfactory formulation seems to be easier. A classical mathematical formulation of the first law of thermodynamics is:

$$\Delta E = Q - W$$

whereby Q, quantity of heat entering the system; ΔE, change in internal energy of the system; W, amount of work done by the system.

It has been much more difficult to formulate the core of the second law of thermodynamics in mathematical terms. A conventional formulation is

$$\eta_{ideal} = \frac{\Delta A}{\Delta Q} = \frac{T_2 - T_1}{T_2}$$

whereby $$\Delta A = (T_2 - T_1) \cdot \frac{\Delta Q}{T_2} = (T_2 - T_1)\,\Delta S$$

which means that the highest possible yield (or, thermal effectiveness) η_{ideal} of a cyclic thermodynamic process (Carnot cycle) is related to the temperature range through which the working fluid drops as it passes through the engine. Note that as T_2 increases or as T_1 decreases, the work output for a given heat input increases. In any case, this is an elegant formulation in symbolic language for a law, which when expressed in English, is as follows: The maximum work that can be obtained from a cyclic thermodynamic process equals the change in temperature multiplied by the change in entropy.

Entropy S is a gauge for the thermodynamic probability W or degree of disorder of the state of a given system. The relationship originally found by Boltzmann in 1866, can be written as

$$S = k \cdot \log W$$

whereby k is the famous Boltzmann's constant. This equation can also be regarded as a kind of formulation of the second law of thermodynamics because the second law states that with all processes occurring in nature the entropy, i.e., the thermodynamic probability W of the state of all bodies or systems involved in the process, will increase. In other words: every *spontaneous* process occurring in nature leads to an increase in thermodynamic probability and thus to an increase in the degree of disorder. This is the real core of the second law of thermodynamics. It implies, e.g., that any increase of order (or, improbability) as it takes place in the development (growth, differentiation, morphogenesis) of every multicellular sytem, must be paid for by an *overproportional* increase of disorder at some other spot of the universe. The source of order (neg-entropy) for living systems is the sun (if we ignore the contribution of artificial light from atomic energy plants).

As we will see later (see Lecture on Physiology and Comparative Biology) there are many strict laws, in particular in comparative biology, in which a formulation in symbolic language is superfluous since a verbal expression permits the highest possible precision. For example, a basic principle of the spermatophytic life cycle: the contents of a fertile embryo sac is homologous to a female gametophyte, can hardly be improved by the introduction of symbolic language. A precise definition of terms will suffice to make this kind of statement unambiguous. This is true for most laws in comparative biology.

We return briefly to the principles of thermodynamics to introduce tentatively the term "prediction." If these principles really apply uniformly throughout the universe (which nobody doubts) they may be axioms in a deductive argument. The resulting statements (equivalent to theorems in a purely axiomatic-deductive system) must be regarded as deduced propositions that are valid for some state of the universe. If these deduced propositions are statements about the future of the system under consideration, we call them "predictions." A theorem deduced from the first and second laws of thermodynamics is that there must be a progressive change of work into heat without a balancing change of heat into work. This means (provided that the universe is a closed system) that the universe is continuously losing order (or, increasing thermodynamic probability). The prediction is that in some remote future the universe will consist of completely disordered matter dispersed throughout space at a uniform, very low temperature. If there is no source of neg-entropy (so far completely unknown to science) somewhere in space or time, the thermodynamic death of the universe is inevitable—this is the prediction derived from the principles of thermodynamics.

We will consider briefly only one further aspect. According to conventional cosmology, the second law yields the only objective gauge for the direction of the arrow of the *physical* time (in the case of a biological time scale, phenomena such as evolution and the directedness of every ontogeny must be considered in addition). The objective gauge for "earlier" or "later" in the physical world is the state of entropy (or, thermodynamic probability) in a closed system that has not yet reached thermodynamic equilibrium, i.e., maximum entropy. Less entropy designates the earlier state, more entropy, the later state. With respect to the universe, the term "time" would become, in a strict logical sense, meaningless after the state of maximum entropy has been reached. The concept of entropy is closely related to the concept of (macroscopic) information. In fact, entropy and information are related by a conservation law that states that the sum of the information and the entropy of a closed system is constant and equal to the system's maximum attainable information or entropy under given boundary conditions. Therefore, a given increase of entropy is always compensated for by a corresponding loss of (macroscopic) information.

The *accuracy* of a prediction depends on the accuracy of the laws implied in the prediction. We recall that laws (as every general proposition) are intellectual constructions based on empirical evidence (singular propositions; objective data) (see Fig. 1). Empirical data can only be obtained with a certain error. The results of measurements are usually distributed in accordance with a normal Gaussian function, which can be characterized by the mean value and the standard deviation, a suitable gauge for the random error involved in taking the measurements. The laws are usually stated without error. In evaluating quantitative predictions we must furthermore keep in mind that the reliability ("precision") of the prediction will be limited by the error involved in the measurements performed to check the prediction. According to the mathematical theory of errors, the random error of a mean can be decreased infinitely (i.e., approaching zero) if the number of measurements becomes larger and larger. Thus, there is no principal but only a practical limitation for the precision of objective data, laws, and numerical predictions within the range of deterministic laws.

In quantitative biology we are always concerned with the properties of populations. Populations of biological things show *variability* with regard to every property or trait. This is due to genetic variation and to the variable influence of environmental factors on the different members of the population. It requires some skill and experience to determine the contribution of genetic and environmental variability to the total variability of the population. The mathematical apparatus to deal with variability is very similar to the one we use to deal with random errors (Gaussian theory). This is fully justified in those cases where the distribution function with regard to variability approximates a normal Gaussian distribution or is at least symmetrical.

A somewhat different situation is encountered in quantum physics. The principle of uncertainty (a universal law)

$$\Delta p \cdot \Delta x = h$$

where Δp = uncertainty in the determination of momentum (m v),
Δx = uncertainty in the determination of position,
h = Planck's constant,

describes the limitation of the degree of predictability in the case of objects with small mass. Electrons, having the least mass of the fundamental particles whose rest mass is different from zero, are most affected, and measurements of their position *and* momentum are most uncertain. Particles with larger mass (protons, mesons, atomic ions, atoms) are much less affected because the principal uncertainty in the measurement of momentum and position decreases with increasing mass. The position and motion of molecules and all larger aggregates are predictable with unmeasurably small uncertainty. Therefore the principle of uncertainty has as a rule no detectable influence on *biological* processes, since we are dealing here with large populations of molecules and ions. However, it cannot be excluded that the behavior of an organism might on occasion depend on the motion of an electron (amplifier model), e.g., in the case of point mutations or in nerve cells. In such cases the principle of uncertainty could indeed influence the shape of the response curve of the organism (2). It is said that the statistical frequency curve of the response would be wide and asymmetric. To my knowledge no experimental evidence has ever been published in biology that would require this kind of interpretation.

Some writers have confronted deterministic laws and probabilistic laws. Deterministic laws, it is said, permit a precise prediction of the particular features of an event whereas probabilistic laws only permit the prediction of particular features of a future event with a certain probability. The laws of quantum mechanics are considered to be probabilistic laws.

The novel situation in quantum physics has arisen from the observational level insofar as seemingly identical experiments turned out not to give identical results. The outcome of any single experiment could no longer be predicted with infinite certainty. It is only the statistical distribution of results of a large number of such experiments that is predictable. It is said that determinism has been replaced by statistical causality. In HEISENBERG'S (12) words: "It is true that in quantum theory we cannot rely on strict causality; but by repeating the experiments many times we can finally derive from the observations statistical distributions, and by repeating such series of experiments we can arrive at objective statements concerning these distributions. This is a standard method in particle physics which may be considered as a natural extension of the traditional method."

However, the epistemologic situation is by no means unambiguous (3). Even the *laws* of quantum physics can be formulated as deterministic laws as shown by the differential equations of wave mechanics. If the laws per se are deterministic, why are the predictions only probabilistic? One could argue that the *states* in quantum mechanics are determined only in probabilistic terms while the laws are deterministic. The principle of uncertainty indeed means that we never can have full information about the physical state of an atomic system at a given instant. It does not imply, however, that the laws governing the transition from one state to the next are not deterministic. Thus on the basis of deterministic (unambiguous) laws a future probability is predicted on the basis of the present probability function. In brief: the probabilistic aspects in quantum mechanics are related to the description of the *state* of the system. The laws that govern the prediction of a future state from the present state may be regarded as deterministic. There is no reason to doubt the universal validity of the principle of causality, because the uncertainty in quantum mechanics concerns the description of the states and not the character of the laws governing the determination of a future state by the present state.

We return to the non plus ultra of science, universal laws, and ask the question whether or not universal laws exist even in biology. It has been emphasized sometimes that even universal biological laws must be considered to be restricted general propositions insofar as they are restricted to living systems and thus probably to our planet with its particular boundary conditions, whereas physical laws are thought to be valid throughout the universe. With this reservation, the answer to the above question is "yes."

Universal laws in biology are principles that are valid for all living systems (but not necessarily for all physical systems). A biological principle that is actually a consequence of the second law of thermodynamics as applied to open systems can be expressed in symbolic language as $\Delta G \neq 0$ whereby G is the symbol for Gibbs free energy. This principle is valid for every open system, living or nonliving. Since *all* living systems *are* open systems, it is valid for all living systems.

As far as the verbal formulation of principles or laws is concerned, there are always two possibilities: a positive formulation (the usual one) and the formulation as negation (4). In the present case, the positive formulation reads as follows: every living system requires the continuous supply of free energy to compensate for the continuous production of entropy. Formulation as negation: there is no living system that is in thermodynamic equilibrium (the formulation in symbolic language represents this negative formulation).

A second example is the principle of compartmentation. Positive formulation: every living system is compartmented. Formulation as negation: there are no living systems that could be considered as homogeneous systems.

A third example is the principle of limited modification. Positive formulation: modifications of a living system by extrinsic (environmental) factors are only possible within genetically determined limits (limits determined by inherited intrinsic factors). Formulation as negation: there is no property (no character) in living systems whose extent would not be strictly limited by intrinsic, genetic factors. This principle is valid also for man. It has never been refuted by facts. Ideologically motivated prejudices against this principle, in particular with regard to psychic or intellectual abilities in man, are scientifically irrelevant.

A fourth example concerns the universality of the genetic code. Positive formulation: the genetic code, i.e., the symbolism and the mechanism by which a sequence of nucleotides is translated into a sequence of amino acids, is the same in every living system from bacteria to man. Formulation as negation: there is no living system that is not subject to the same common genetic code.

Next we study the particularly instructive example of a statement that was considered to be a principle but was refuted experimentally. Kornberg described an enzyme that replicates DNA in the test tube (DNA-replicase or Kornberg's enzyme). This enzyme was considered to be *the* DNA replicating enzyme of *the* cell. Formulation as negation: There is no cell that does not depend on the Kornberg enzyme for DNA replication. The experimental refutation of this statement was published in 1969 by Cairns. Among thousands of mutant strains of bacteria he found a mutant that could replicate DNA and multiply at a high rate without having the Kornberg enzyme. This showed that the Kornberg enzyme is not essential for DNA *replication* in vivo. After Cairn's discovery, the foregoing principle was considered to be refuted. The research field of DNA replication was thus opened for new discoveries.

A principle of great importance in science *and in epistemology* will finally be considered. Positive formulation: All physical laws (= laws of physics) may be applied to living systems (or, are valid in biology). Formulation as negation: There is no physical law that cannot be used in the construction of theories in biology.

This principle is *not* reversible. There are laws in biology that cannot be used in physics. This has to do with the phenomenon of genetic evolution, which has been an event *specific* for *living* systems. Evolution has led to systems whose complexity is far beyond the complexity (not size!) of any physical system.

Most theories consist not only of general (in a strict sense, theoretical) laws comparable to axioms but also of law-like statements of a more observational nature that can be directly tested any time by observation. We may call these statements "empirical laws". I think that it is the empirical laws—mostly "simple" laws of limited scope—which make up the stable, really reliable parts of the network of general propositions that constitute a theory. If a prediction is made from an empirical law our confidence in the validity of the prediction is very strong since it is a case of inference from undoubtable evidence to prediction, by-passing the more risky universal generalizations of the theory. It is the empirical law on which technology is predominantly based.

Empirical laws are often quite simple, i.e., they contain only a few arbitrary parameters (5). Simplicity as a rule includes an element of esthetic appreciation; scientists sometimes use the term "elegant law" in order to express their sentiments. In any case, logical simplicity is not only characteristic for the highly theoretical universal laws, for the principles; rather it is preferred whenever possible even in the realm of empirical laws (5).

Empirical laws may have different natures. We briefly describe two classes: process laws and coexistence laws. A process law permits the prediction (or retrodiction) of future (or past) states of a system, provided the values for the relevant variables at any one time are at hand. A coexistence law describes the *contemperaneous* existence of properties of a system.

Examples of Process Laws. The law of radioactive decay is a typical example of an empirical process law in physics. It reads in verbal language

(5): the rate of radioactive decay of a substance is independent of its temperature, pressure, or state of chemical combination. In symbolic language it can be expressed by the equation

$$N_t = N_0 \cdot e^{-k \cdot t}$$

where k is the decay rate constant, which does not depend on temperature, pressure, or state of chemical combination.

A corresponding empirical process law in biology is the law of exponential growth. In verbal language: the growth rate of a system is proportional to the amount of the system already present. In formal language: $\frac{dN}{dt} = k \cdot N$. This differential equation has the solution

$$N_t = N_0 \cdot e^{k \cdot t}$$

where k is the rate constant for growth. In this case the numerical value of the rate constant *does* depend on environmental factors such as temperature.

Examples of Coexistence Laws. The gas law is a typical example of an empirical coexistence law in physics. In verbal language: the volume of a given mass of an (ideal) gas varies inversely with the pressure to which it is subject at constant temperature. In symbolic language it can be expressed by the equation

$$pV = \text{constant} \qquad (= RT)$$

This gas law is a fine example of the "correction" (= adjustment) of an empirical law under the influence of a theory, i.e., under the influence of an integrated network of concepts including statements about intermolecular attractive forces and sizes of molecules. The gas law in its original form $P \cdot V = \text{constant}$ describes the behavior of real gases only to a first approximation or, formulated negatively, the law was refuted in its original simple form by refined measurements. The law was "saved," however, in principle by van der Waals who extended it to

$$(P + \alpha/V^2)(V - \beta) = R \cdot T$$

where the parameter α relates to intermolecular attractive forces and β relates to the actual size of the molecules.

A corresponding empirical coexistence law in biology is the law describing the state of water in a cell. It can be expressed in verbal language by the statement that the water potential of a cell is determined by the osmotic potential of the cell sap, which is counteracted by the pressure potential and matrix potential. In symbolic language:

$$\psi = \pi^* - P - \tau$$

where ψ = water potential; π^* = osmotic potential; P = pressure potential; τ = matrix potential. We have already mentioned that the formulation of this law depends on a particular construct "cell," which emphasizes the elastic cell wall, the semipermeability of the thin marginal layer of the protoplast, and the osmotic value of the contents of the "vacuole" (cell sap).

A law must not *necessarily* imply a causal relationship between two (or more) variables. This becomes obvious when we consider correlations that can be classified as coexistence laws. If the relationship between x and y is characterized by a correlation coefficient close to +1.00, a perfect prediction of y is possible if x is known, even though we must not interpret, as a matter of course, a high correlation in terms of a causal relationship. Take a weather forecast. A low reading of the barometer permits the prediction of a storm or at least of some kind of bad weather. It is obvious that the barometer reading and the storm are closely correlated but not *causally* connected.

A special class of coexistence laws in biology describes allometric growth. If we plot the increase in length (x) of an organ vs. the increase in width (y) on a double logarithmic scale and obtain a straight line, the situation is referred to as allometric growth. Allometric growth indicates that growth in one direction is strictly related to growth in another direction. In other words, there must be a superior control system for growth in different directions.

The empirical allometric equation $\log y = a \log x + \log b$ (or $y = b \cdot x^a$) is the result of the integration of the differential equation

$$\frac{\frac{dy}{dt} \cdot \frac{1}{y}}{\frac{dx}{dt} \cdot \frac{1}{x}} = a \quad \text{or} \quad \frac{dy}{y} = a\,\frac{dx}{x}$$

which means that the relative growth rate in one dimension is related to the relative growth rate in another dimension by the constant factor a.

The Hardy-Weinberg law is generally considered to be a prototype of a coexistence law in biology despite the fact that it suffers from two shortcomings: (1) it is only valid under restrictive boundary conditions and (2) it refers to theoretical entities (genes, alleles) without considering the bridge principles between the genotypic and the phenotypic levels. The Hardy-Weinberg law is usually expressed by the binomial equation

$$p^2 + 2pq + q^2 = 1$$

whereby p^2 represents the frequency of one homozygote A_1A_1, q^2 represents the frequency of the other homozygote A_2A_2, and $2pq$ represents the frequency of the heterozygotes A_1A_2 in a given population.

In verbal language: in a (infinitely) large, sexually reproducing population in which males and females mate at random with regard to the genes in question ("panmictic population") and produce equal numbers of equally fertile offspring, the ratio between two alleles will remain p:q, supposing that this ratio was once established at some point in the history of the population and further supposing that there are no outside influences and no mutations to and from the two alleles involved (A_1, A_2).

The particular value of the law lies in the prediction that if the gene frequencies predicted by the law did not hold from generation to generation in natural populations, something in the boundary conditions outlined above must have changed; some factor must have operated within (e.g., mutations) or on the population (e.g., a selection pressure). As RUSE (6) has emphasized, the Hardy-Weinberg law can be derived from Mendel's first law and from the boundary conditions just mentioned. Thus, the Hardy-Weinberg law cannot be considered to be an empirical law; rather it constitutes a case of derivation of a particular law from a more general law in biology.

As an example of an empirical coexistence law that is as important in physics (and physical technology) as it is in biology, we refer to Poiseuille's law of viscous volume flow. In symbolic language:

$$V = \frac{\pi(P_1 - P_2)}{8\eta L} R^4.$$

In verbal language: the rate of flow of a given liquid through a tube is proportional to the pressure drop across the ends of the tube. The law has usually been expressed as a diagram (Fig. 2).

Figure 2. Rate of flow of a given liquid through a tube is directly proportional to pressure drop across ends of tube ($P_1 - P_2$). This empirical law (Poiseuille's law of viscous flow) holds for glycerol and for water, but not for solutions of certain polymers [adapted from (5)]

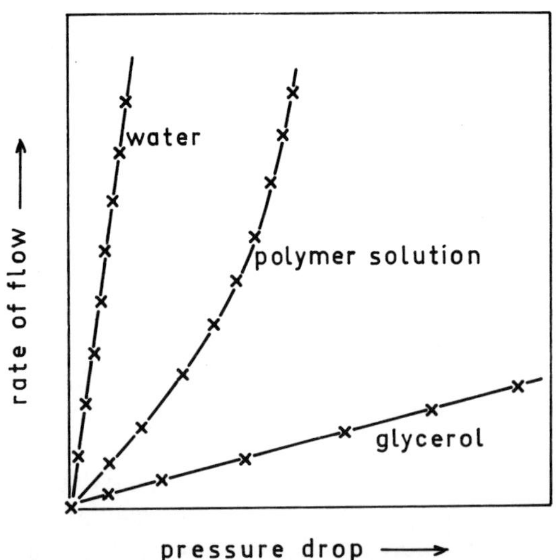

The diagram shows that this law is a limited law (or restricted law) since it holds indeed for glycerol and for water but not for solutions of certain polymers. The reason for the failure of the simple law in the case of certain polymers is that these molecules can no longer be regarded as tiny, independent, hard spheres without significant intermolecular forces. Rather we must consider them as long thread-like molecules. This has the consequence (again derived from concepts of the theory of which the empirical law is an integral part) that the viscosity varies with the flow rate. We may learn from this example that, usually, the scope of simple empirical laws is restricted. Simplicity (and elegance) is obtained at the expense of generality or accuracy. The manner in which the scientist proceeds if a simple empirical law fails is nicely illustrated by Poiseuille's law: The deviation of the polymer solution from the law is "explained" by referring to another law that describes the effect of the flow process on the molecular configuration. In the case of water or glycerol this effect can be ignored since these molecules "behave" as hard, rigid spheres unaffected by the flow process (5).

As a rule empirical laws are valid over a limited range of boundary conditions. It is a convention in science to test the laws only over a limited range of variables. Therefore the empirical laws are valid only over this range. But *within* this range these laws are very safe and thus make up a kind of backbone for the general propositions of a theory. Furthermore, empirical laws form a reliable basis for technology, i.e., application of scientific laws in a scheme of teleologic action (see Fig. 11).

Returning briefly to Poiseuille's law, we ask a question of utmost importance for the structure and motivation of science. When, i.e., under what circumstances, do we really want to "explain" something in science? The answer is (in science as well as in daily life): we start from problems; we want an explanation when we meet some *unexpected* state of affairs; we try to explain why something we did *not* expect could happen. In more precise terms: we are forced by the very nature of our mind to find an explanation if our conscious or unconscious expectation and the observation we make *deviate* from each other. We want to explain the discrepancy. How? We search for "new" laws, i.e., laws so far not used in the theory, or we try to explain the discrepancy between prediction (expectation) and the actual event by a change in the antecedents (initial and/or boundary conditions). For the practicing scientist, it is obvious that a major force driving the progress of science is the "surprise effect," i.e., the emotional response that arises if an experimental result or an observation turns out to be *different* from what we expect. The strong desire to explain an *unexpected* event is characteristic for our daily life as well as for science. Also in this regard science is clearly nothing more than a highly disciplined conscious and intellectually controlled extrapolation from our normal behavior. In the case of Poiseuille's law the strong deviation of the polymer solution from the law *had* to be explained. It has been explained by a new law, which describes the effect of the flow process on the molecular configuration, an effect that does not come into play as long as we deal with small molecules, such as water or glycerol, which can be considered to be tiny, hard spheres whose shape is independent of flow rate.

To summarize this aspect: Our inborn passion to explain and to earn recognition for right explanation is possibly the most important root of science. In course of the genetic evolution of man, this passion originated under the selection pressure in favor of solving problems *before* the unsolved problem eliminates the genotype. This implies that truth will help to survive. We will later see that this conviction is part of the intrinsic value system of science.

We now turn to a systematic analysis of the process of explanation. Most scientists simply take it on trust that they are familiar (as a matter of course) with the logical structure of an explanation. In practice, however, the "explanations" advanced by scientists differ considerably from each other. It has been the merit of several philosophers of science, in particular HEMPEL (7), to have developed a model (sometimes referred to as Hempel-Oppenheim model) that covers the logical structure of scientific explanation as well as of a scientific prediction (Fig. 3).

There are three elements in the Hempel-Oppenheim model: one or more general propositions ("laws"; L_1, L_2 --- L_x), a set of statements about particular circumstances (boundary conditions and/or initial conditions; C_1, C_2 - - - C_x) and a statement of the event to be inferred from the former two sets of statements (E or P). In the case of an explanation, the general propositions in conjunction with the boundary or initial conditions form the explanans. The statement of the event being explained is the explanandum. In the case of a prediction, the general propositions in conjunction with the boundary and/or initial conditions form the premises. The statement of the event being inferred is the prediction. The validity of the principle of causality as the most basic principle in science is always implied.

According to this scheme (Fig. 3), explanation and prediction in

Figure 3. Logical structure of explanation and prediction according to HEMPEL and OPPENHEIM (7). Symbols are explained in text. Note with C_1 - - - C_X that boundary conditions—in addition to initial conditions—must be specified, if system under study is not isolated. Symmetry thesis applies only when explanandum denotes individual event

L_1, L_2 --- L_x
C_1, C_2 --- C_x
} explanans (premises)

E (P) } explanandum (prediction)

} explanation (prediction)

science refer back to some kind of "law" (some kind of general proposition). Since there is a wide variety of general propositions there is also a wide diversity of "explanations," from the vague "explanation" of a headache or the flu to the quantitative explanation of a solar eclipse. The strength of prediction and explanation depends on the very nature of the law(s) implied. Scientific explanations (or predictions) should be based, of course, on general propositions with greatest possible confidence; however, it would be a misunderstanding of the nature of science if one would postulate (as some rigorous philosophers have tried) to permit only universal laws in the Hempel–Oppenheim scheme.

In many cases, one wants to explain a *law*, i.e., the explanandum is itself a law rather than an event. In this case, one has to show by deductive inference that one law follows from another law whereby the Cs in the Hempel–Oppenheim scheme do not play a role. As an example, the empirical process law $N_t = N_0 \cdot e^{-k \cdot t}$ follows from the more general law that there is no cooperativity between the members of a population of radioactive atoms or—in other words—that the radioactive degradation is a random process.

In the Hempel–Oppenheim model there is obviously some asymmetry since an *explanation* implies that the event has already occurred (i.e., the element E is real) while in the case of a prediction the real occurrence of the actual event is not automatically inferred (1). P *can* occur but it must not occur. This asymmetry is a trivial thing for a scientist who has ever taught about the thermodynamics and the actual reaction kinetics in a case where activation energy or enzymes come into play. If we consider only the prediction of a *potential* event, the explanation–prediction model looks symmetrical.

There have been many objections to the Hempel–Oppenheim model, some of them obviously justified. However, the central core of the model is clearly valid: scientific explanation is an act of deductive inference, and explanation and prediction are structurally very similar. However, the symmetry thesis should not be maintained dogmatically if obvious difficulties arise such as in the explanation and prediction of rare or unique events, e.g., new point mutations. As an example, X-ray-induced point mutations are probably always quantum events. They occur at the level of single molecules and they belong to the realm of microphysics where the principle of uncertainty comes into play. This excludes a precise *prediction* and also a control of the *individual* event, whereas an explanation of a mutation (once it has occurred) is not impaired.

A particularly interesting and intensely debated case in this connection is offered by the so-called historical laws of evolution. Every biologist knows that the "laws" of evolution (= phylogeny) and the laws of evolutionary mechanism must be kept separate. The laws of evolution are hardly more than descriptive generalizations of limited scope and with the risk of exceptions, such as Cope's rule ("in most evolutionary lines the body size increases") or Dollo's "law" ("evolving groups do not retrace the stages of their evolutionary development to return to an earlier state"); in brief: "evolution is irreversible."

The laws of evolutionary mechanism are, at the present state of our knowledge, mainly the statistical laws of population genetics, i.e., in essence, probabilistic statements on random mutation, recombination, and adaptive selection toward fitness. These laws of population genetics are by no means historical laws since their validity does not depend on their application to the case history of phylogeny. In my opinion the historical aspect in biology has been largely overestimated in the past. Certainly, all living systems are products of a historical process (evolution = phylogeny) and must be treated theoretically before this background. We will return to this aspect in the Lecture on Epistemology and Evolution. However, would a statement such as "living organisms are products of history and must be treated differently from inanimate objects" (8) really be justified? In fact, the second law of thermodynamics is not only a process law but also a historical law since it makes firm statements about the unidirectional development of the universe. Not only species have histories, but also the universe, galaxies, stars, the solar system, and the earth. The fact that all the planets in our solar system move in more or less similar planes that are not far apart and in the same direction around the sun is clearly accidental, i.e., it must be due to a peculiar historical event. Of course, the gravitational laws have no history (the same is true for the laws of population genetics; scientific laws have been held to be timeless, at least since the adoption of uniformitarianism in geology); the solar system, however, obviously has history. For this reason Kepler's "laws" must be regarded as singular propositions (basic judgements) rather than as laws.

Sometimes it is indeed difficult even in physics to decide whether a particular statement has the characteristics of a basic judgement (singular proposition) or whether we deal with a law-like structure. A good case to illustrate this important point is Kepler's laws. The three laws postulated by Kepler are:

1. Each planet moves in an elliptical orbit around the sun, and one focus of the ellipse is located at the sun.
2. The line from the planet to the sun sweeps out equal areas in equal times.
3. The length of time needed for a planet to revolve once around the sun is called its period. The square of the period is proportional to the cube of the planet's greatest distance from the sun.

Some modern philosophers of science (e.g., Scheibe) point out that Kepler's laws are actually not laws but basic judgements (singular propositions), which follow from the gravitational laws and the specific boundary conditions of the solar system, and which can only be understood historically.

A similar situation seems to exist in the theory of evolution. We mentioned previously that the laws of the evolutionary mechanism are predominantly the probabilistic laws of population genetics. MARY WILLIAMS (9) has performed an axiomatization of evolutionary theory insofar as she succeeded "to express Darwin's theory of evolution as a deductive system in which a few fundamental principles of the theory are used as axioms from which the remainder of the principles of the theory can be deductively derived." Thus, WILLIAMS has derived laws from a given set of axioms, laws that govern the process of evolution and permit predictions about trends and patterns of evolution. This is probably the maximum that can be achieved in a theory of this kind, since the actual progress of evolution of any particular taxon will not only be determined by the laws (or principles) of evolution but also by unique historical configurations of circumstances (initial and/or boundary conditions). This excludes any *precise* prediction of the future course of evolution since these boundary conditions are not precisely known. Some of them always lie in the future themselves.

This again is not a peculiar feature of *evolutionary* prediction. Rather, this is true whenever the initial and/or boundary conditions are at least partly in the future. Under these circumstances only if–then predictions are possible. I have already described that all predictions made by scientists in any political context are only if–then predictions, since at least some of the more political boundary conditions, in particular those that require decisions, cannot be taken into account in any binding way.

The reconstruction of phylogeny is a historical venture. It must be *compatible* with the laws of evolutionary mechanism. But the major material for the reconstruction of phylogeny remains the fossils and the laws of comparative biology.

According to the Hempel–Oppenheim scheme (Fig. 3), explanation and prediction in science refer back to some kind of "law" (or to laws) and to a given set of boundary conditions. This scheme describes the structure of *causal* explanations in a satisfactory manner. We observe an event or state of affairs and we want to explain as precisely and as quantitatively as possible how it could come about. The Hempel–Oppenheim scheme also describes the structure of a scientific experiment. Every experiment consists of two parts: a precise, quantitative prediction and an experimental test to check the validity of the prediction. It is obvious from the scheme that an experiment is only possible if the boundary conditions can be strictly defined and maintained. An example that stands for every good experiment in science is the following (5): Galileo discovered in his famous experiments of 1589–1591 that various heavy bodies of different weights reached the ground at the same time if dropped from the Leaning Tower of Pisa. He concluded that the free fall of all heavy bodies does not depend upon their mass and is governed by the same law ($X = \frac{1}{2}g \cdot t^2$). On this basis one can predict that even a feather and a ball of lead will fall at the same rate if friction, air movements, collision with other particles, etc., are completely excluded. This is the case in a vacuum. The experiment confirms this prediction. Galileo could not make this experiment with the feather and the ball of lead on the Tower of Pisa because the complex and changing boundary conditions in the open air affected the very light body in a much more severe manner than the heavy bodies he used. Only with the heavy bodies could he approximate *equal* boundary conditions (Fig. 4).

It is sometimes stated that in order to predict the rate of free fall in a vacuum one only needs one law ($X = \frac{1}{2} g \cdot t^2$) and the numerical value of g as a boundary condition ($g = 9.81\ \mathrm{m} \cdot \mathrm{s}^{-2}$). This is not correct insofar as it is an *empirical* law rather than a priori knowledge that the ratio between mass and weight is constant for every kind of body, even for such different ones as a ball of lead and a feather. An old jest question is the following: what is heavier, a pound of lead or a pound of feathers? It is amazing how many people will quickly give the wrong answer. In addition, the validity of the principle of causality is implied in the predictive scheme as a matter of course.

The design of an experiment requires experience, as a rule, highly specialized experience. In a scientific communication of experimental results the boundary conditions must be described so precisely that *potentially* the experiment can be repeated by every competent person in every other equally equipped laboratory. This is the basis for *objective*, experimental data, for "common knowledge." The notion of the "repeatable, controlled experiment" can hardly be overestimated. Only repeatable experiments contribute to the body of common knowledge of a scientific community.

While it is agreed among scientists that precise measurements and the use of numbers are essentials for every kind of causal research, the *logical* structure of causal research is often misunderstood. I will describe the principle of causality in the forthcoming lecture. However before we

Figure 4. Application of the Hempel-Oppenheim model to Galileo's experiments with falling masses. (It does not matter whether Galileo in fact performed these experiments during his stay in Pisa between 1589 and 1591. In any case, Galileo must be regarded as the "Father" of the quantitative physical experiment. In addition, he found the correct formulation of the law of the free fall, probably by the middle of the year 1604)

$$L_1\colon x = \tfrac{1}{2}\, gt^2,\ L_2\colon \frac{\text{mass}}{\text{weight}} = \text{constant},\ L_3\colon \text{causality principle}$$

$$C_1\colon \text{vacuum} \quad C_2\colon g = 9.81\ \mathrm{ms}^{-2}$$

P : Within 10 s a heavy body will fall 490.5 m

can turn to the principle of causality it is necessary to vindicate two kinds of explanations that are peculiar to biology and therefore unfamiliar to the physicist and to the physics-oriented philosopher: *functional* explanation and *teleological* explanation. These kinds of explanation play a major role in descriptive and comparative biology, and the question arises whether or not they can be vindicated.

Functional accounts are explanatory insofar as they state what role a part plays within the whole. "Part" can be any structure, any feature, any process, any piece of behavior. The "whole" can be an organism or a group of organisms ("The function of chlorophyll is to absorb light quanta for photosynthesis"; "The function of flower anthocyanin is to attract pollinating animals"). The notion of function is originally extrinsic to the system. It is imposed upon the system in order to make the system and its parts comprehensible. Purposiveness is a category we impose on living systems as a matter of course. As KANT (10) puts it: "We only read this conception into the facts as a guide to judgement in its reflection upon the products of nature." I assume that the "automatic" application of purposiveness to living systems is part of our inherited foreknowledge about nature (see Lecture on Epistemology and Evolution). While Kant considered purposiveness a heuristic maxim, we are convinced today that purposiveness is a constitutive intrinsic category. However, since we believe in Darwinian evolution as a paradigm, we insist that a functional explanation in biology does not imply finalism as a metaphysical doctrine. It is *not* implied in the notion of purposiveness as a constitutive intrinsic category of living things that the function of the parts of the whole is determined by ultimate goals or purposes in a metaphysical sense. We simply assume whenever we advance a functional explanation in biology that the whole system (a plant, an animal, a population, an ecosystem) has been genetically optimized during evolution, i.e., adapted to reproductive fitness, and this implies that there are no useless parts. However, if there are obviously useless, imperfect, or poorly functioning parts, they may be *explained* in terms of rudimentary organs. This kind of explanation refers back to the theory of evolution that permits an explanation of poor performance or efficiency of an organ in terms of the peculiar exigencies of its phylogenetic origin (1).

Even the molecular biologist assumes as a matter of course that living systems are more or less functionally optimized systems. A statement by the microbiologist MAALØE (11) ". . . the reason that a bacterium operates like a well-organized factory is . . . that, in the course of evolution, the bacterial cell has "learned," in a given growth condition, never to produce more ribosomes than it can put to use with very high efficiency."

As a rule, functional and causal explanations are complementary. The biologist resorts to functional explanations not because a causal or mechanistic account of a life process could not always be achieved in principle, but because functional explanations are demanded by the very nature of biological systems and by the type of understanding we seek.

Functional explanations may sometimes play a valuable role as a guide for *causal* explanations in physiology. However, there are serious difficulties and even traps. For example: (1) Many parts or elements in living systems do have *multiple* functions. This is true for organs (e.g., a root) as well as for molecules (e.g., we find anthocyanins in seedlings, flowers, and fruits. Furthermore we find them in the autumn leaves where they have no obvious function (2). Some features of organisms are not useful by themselves. They have arisen as concomitant or incidental consequences of other features that *are* adaptive. The foliage coloration in the fall is an obvious example. In some cases such features have *become* adaptive at a later stage. For example, the sound of the beating heart has become useful for the healthy mother–child relationship in man. (3) The function of a part within the whole usually requires a fully developed system equipped with homeostasis, e.g., a fully grown, green leaf in the case of optimum photosynthesis. The developmental process that leads to the functional state of an organ may not be accessible to a functional explanation.

At this point *teleological* explanations in a strict sense come into play, e.g., statements such as "The reason for the existence of proplastids in plant cells is that these particles provide the starting point for the development of chloroplasts" or: "The reason for the existence of 'eyes' on a potato tuber is that these structures permit the rapid formation of new sprouts." It is not implied in such statements, of course, that the proplastid or the dormant bud on the tuber tend "consciously" toward their specified and functional end state. Teleological explanations imply only that the existence of a structure or a part can be "understood" by referring to a *future* functional state (end state). There is no "final cause" in the Aristotelian metaphysical sense involved in this kind of a teleological explanation; the end state is causally and in general also temporarily posterior (1). The only assumption we must make (as a rule by implication) is that not only the functional end state but also the processes leading to the functional state are genetically optimized in the sense that they are evolutionary adaptations that contribute to the reproductive fitness of the species. With regard to the developmental geneticist who aims at *causal* explanations of developmental events, teleological explanations may provide a valuable and sometimes even indispensable basis for argument, hypothesis, and experiment.

Teleological processes of a higher and qualitatively different level [sometimes referred to as external teleology (1)] are those where the end state or goal is *consciously* anticipated by the agent. This is aimed, purposeful, and deliberate activity as it occurs (according to our subjective experience) in man and probably also in higher animals. While purposeful activity is thus part of nature and a legitimate subject of *biological* inquiry, it seems that it is restricted to some highly developed living systems. We will return to the structure of teleological *action* after we have dealt with the principle of causality and its consequences.

I mentioned previously that functional and teleologic explanations are not being used in physics. However, functional explanations play a predominant role when we deal with *technological* systems. Every system which is the result of creative, purposeful *activity* by man can be treated intellectually in terms of functional explanations since we may assume that every part of the whole serves a defined and distinct purpose.

The fact that precisely the same type of functional explanation may be applied to biological as well as to technological systems may *not* lead to the conclusion that biological systems are as well the result of teleological purposeful action of some extrinsic agent ("creator," "life force," "élan vital"). Technological systems are clearly the result of teleological thought and action; living systems, however, are the product of evolution. The *causal* explanation of the purposefulness of the products of evolution does not require any teleological agent. The present theory of evolutionary mechanism permits a satisfactory explanation without the assumption of any extrinsic agent (see Lecture on Epistemology and Evolution).

6th Lecture

The Principle of Causality

Causality to a scientist conveys the general proposition that every event has a cause. Causality is considered to be a principle, i.e., a universal law or even a conditio sine qua non of science (1). I will describe mainly the structure of the principle of causality as it is presently used in physiology, i.e., in *biological* research, which aims at the *causal* explanation of an event. Since the structure of causal research is often misunderstood even by experienced scientists it might be worthwhile to describe it explicitly. Causal research in biology should actually be called factorial or factor research since we are not able to study the relationship between cause and event but only the relationship between variable factors and events. This is, as we will see, not only a semantic problem.

Formulation of the principle of causality requires the notion of determination and a time directedness. The principle can be formulated as an if-then proposition (Fig. 5): If x distinct factors (F_1, F_2 . . . F_x) determine the state a and if a′ follows from a with time, then the general proposition is that if the state a (determined by x factors) is given (cause), the state a′ (effect) will *always* follow. A formulation of the principle of causality as a negation is: There are no indeterminate events.

Whether the principle of causality is true is a difficult question to decide, but it is certainly assumed to be true by every practicing scientist. Otherwise science, in particular prediction, explanation, and teleological action would not be possible. We mentioned previously that even principles, i.e., general propositions of the highest possible rank, are subject to possible refutation by experience. As an example, the "law of parity" in physics, which was thought to be very fundamental, was disproved, for weak interactions in recent years. The principle of causality, however, cannot be tested in this manner since it is *assumed* to be valid in everything we do in science. In every experiment designed to refute the principle—there are no indeterminate events—we must *imply the validity* of this principle.

Therefore the principle of causality cannot be treated on the same level as the other universal laws. The principle of causality is an indispensable prerequisite of science, irrespective of whether we classify it as a universal law or as a metaphysical postulate.

The principle of causality is indeed a good example of the fact that the scientific venture to explain the world is based on a set of a priori categories of the human mind whose validity cannot be investigated with strictly scientific means. However, these categories are neither cultural beliefs nor are they randomly adopted intellectual creations. Rather, they are part of our a priori knowledge about the world and thus part of our *genetic* heritage rather than part of our *cultural* heritage. We call these categories, which are part of our genetic information and on which we depend for any orientation *in* and intellectual organization *of* the world, *constitutive* categories. Without a priori constitutive categories the world would not be intelligible. This aspect will be dealt with further in the Lecture on Epistemology and Evolution.

Using the model in Figure 5 we treat the structure of biologic causal research, which must always be regarded as "factor analysis," rather than as causal analysis. In "causal" research we cannot do more than to vary one, two, or (rarely) more factors in an experiment and to record the effect. Those of the x factors that we vary in the experiment are called variable factors or briefly "variables." The logical structure (mode) of "causal" analysis can be illustrated as follows: With F_1 (F_1, F_2, respectively) we designate the experimental variables. The state of the system *without* the variables we designate by A, the resulting effect by A′. The *change* of the system under the influence of F_1, (or F_1, F_2, respectively) we designate with ΔA, the changes of the effect with $\Delta A'$. By this approach it is possible to

Figure 5. Formulation of principle of causality to indicate use of this principle in "causal" research in biology [after (2)]

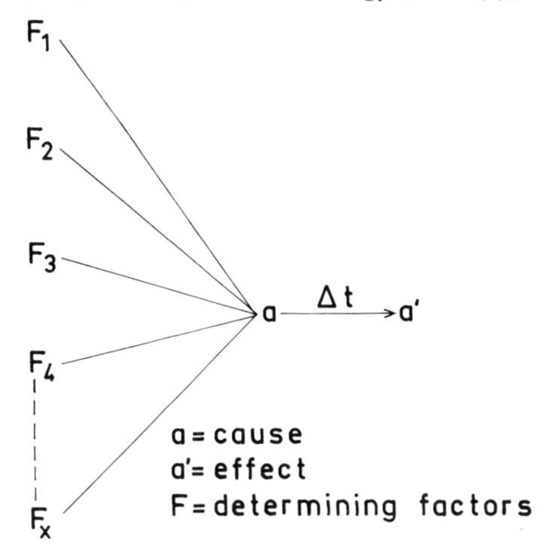

pg 78
determine $\Delta A'$ as a function of F_1 (or F_1, F_2) quantitatively even though we do not know (x − 1) or (x − 2) of those factors that determine a and bring forth a′ as illustrated in Figure 5. We recall that a is determined by x factors and that a′ follows necessarily if a is determined. The conjecture of a *causal* analysis would imply that we are in a position to describe *how* the x factors ($F_1 \ldots F_X$) determine a and bring about a′. Even with the most simple living system such as bacterial cells we are far from being able to perform a full-scale causal analysis. Therefore the physiologist will restrict himself to a factor analysis whereby $A + \Delta A = a$ and $A' + \Delta A' = a'$. In practice, $\Delta A'$ will be measured as a function of the quantity of the experimental variable

$$\Delta A' = f(\Delta F).$$

Since the physiologist is predominantly interested in the factorial analysis of *processes*, he will usually try to determine how fast the change of the variable will lead to a change of the effect

$$\frac{\Delta A'}{\Delta t} = f\left(\frac{\Delta F}{\Delta t}\right).$$

This kind of description of the principle of causality (3) accounts for the usual way of thinking of events as being *made* to occur by the addition of a factor as well as for the more scientific way of asking what factors *prevent* an event from happening.

In an attempt to illustrate the formal model of the principle of causality we consider the case of a single factor analysis in genetics. To characterize living systems, the scientist uses traits (characteristics, characters). These terms designate such properties or abilities of living systems that can be *measured*. The sum of the traits is called the phenotype.

As an example, we chose the trait "anthocyanin." The red plant pigment anthocyanin can easily be detected and be measured quantitatively. Since the formation of anthocyanin is a luxury function of the cell, i.e., not required for the existence of a cell per se but only relevant for the reproductive fitness of the whole organism, this trait has been used extensively in genetics. Under *experimental* conditions it is irrelevant for the well-being of a plant whether anthocyanin is being synthesized or not. Comparing Figure 6 with the model (Fig. 5) it becomes obvious that those x genes that contribute to the trait anthocyanin must be identified with the system *a* whereas the trait proper (the appearance of anthocyanin) must be identified with the effect a′. Now we choose one of these x genes as a variable, say $gene_4$. This particular $gene_4$ is equivalent to the variable factor F_1 in the general model. Without $gene_4$ the trait will not develop at all even though all other factors are available. The law-like relationship between the amount ("dose") of $gene_4$ and the amount of the trait ("effect") is indicated in Figure 7. In the case of intermediary inheritance in a diploid system, $gene_4$ is the rate-limiting factor whereas with recessive-dominant inheritance the gene dose 1 is already saturating. This implies that some

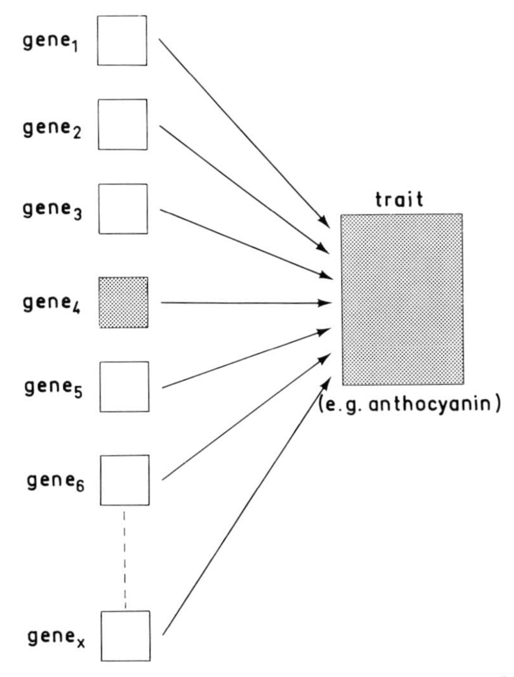

Figure 6. This sketch is supposed to illustrate *formal* gene-trait relationship and "logic" of single-factor analysis. "Trait" is used in sense of classical genetics. E.g., red coloration of petals by anthocyanin is considered a trait [after (2)]

Figure 7. Quantitative relationships between gene dose (factor dose) and amount of trait (effect) in a diploid system producing anthocyanin. Intermediary inheritance: amount of anthocyanin is proportional to gene dose. Recessive-dominant inheritance: gene dose 1 (of $gene_4$ in Fig. 6) saturates system. With increasing gene dose other factors limit production of anthocyanin [after (2)]

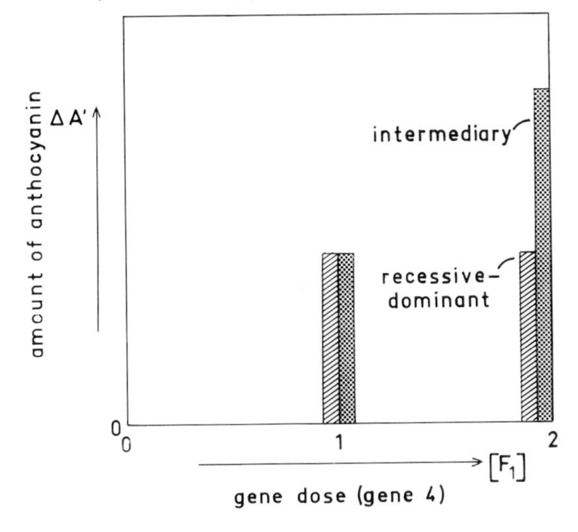

other factor(s) of the x factors become limiting with regard to the quantitative expression of the trait anthocyanin as soon as the gene dose 1 of $gene_4$ is available. Most of classical genetics is based on this model. It is obvious that a statement such as "$gene_4$ causes anthocyanin synthesis" is not an appropriate description of the actual situation. Anthocyanin synthesis is caused by the total set of x genes. However, the statement "the lack of appearance of the trait anthocyanin is caused by the lack of $gene_4$" is logically correct and should thus be preferred.

In many systems anthocyanin synthesis does not occur even though all genes required for anthocyanin synthesis are available in the system. An external factor, light, is required in addition. In a precisely investigated system, the mustard seedling *(Sinapis alba)*, it could be shown that light acts exclusively through the formation of an effector molecule, active phytochrome, P_{fr}, from a physiologically inactive precursor. It was further shown that the "dose-response function" (amount of anthocyanin as a function of the amount of the effector molecule P_{fr}) is linear over a considerable range and extrapolates through point zero (Fig. 8). This means that the relationship between $\Delta A'$ (amount of anthocyanin) and the dose of the rate-limiting factor (F_1 in the general model) can simply be described by the equation

$$[\text{amount of anthocyanin}] = c \cdot [P_{fr}]$$

Figure 8. Quantitative relationship between dose of effector molecule P_{fr} (made by brief light pulse) and amount of anthocyanin produced within 24 h by given dose of P_{fr}. System: mustard seedling *(Sinapis alba)*. The dose-effect curve is linear in neighborhood of zero and extrapolates through point zero [after (4)]

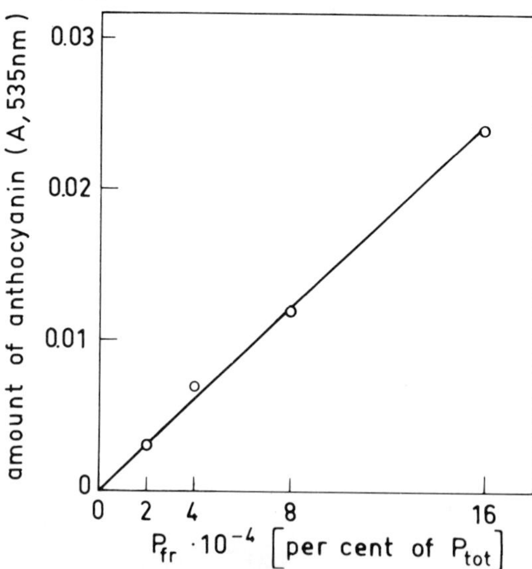

where *c* is a constant, specific for the system. This implies not only that the other factors are not rate-limiting; this result also implies that there is no interaction between the variable and the set of the other factors. This is a prerequisite for the analysis of those processes that are set in motion by the factor P_{fr} and eventually lead to the appearance of a certain amount of anthocyanin. Thus the logical requirements for successful factor analysis are met.

Sheer ignorance of the logical structure of "causal" research in physiology has created much confusion. As an example, we shall consider briefly longitudinal growth of a plant stem or coleoptile section. According to our model (Fig. 5) growth rate $\left(\frac{\Delta L}{\Delta t}\right)$ will be determined by x factors (growth factors). The collective of the growth factors consists of two subcollectives: intrinsic growth factors (intrinsic with regard to the growing system) and extrinsic growth factors (extrinsic with regard to the growing system). Light, e.g., is an extrinsic growth factor. Some of these growth factors will be available in nonlimiting amounts. As a rule the growth rate will be limited by only one or a few factors. Those factors that are actually rate-limiting for the growth response, we call "growth regulators." It is obvious from the model in Figure 5 that any of the x factors are required for growth (i.e., are part of the *cause* for the growth *effect)* even though only a few of the x factors are actually limiting. It is possible to make a growth regulator out of any growth factor. As an example, the ATP pool is, of course, a growth factor but under normal conditions not a growth regulator. By impairing ATP production in a given system, the ATP pool can easily be made rate-limiting. By impairing the biosynthesis of gibberellin, this "hormone" can become a growth regulator even though it may not limit growth under normal conditions, etc.

If we work with isolated segments instead of the intact plant, there is a big chance that at least some serious changes occur at the level of the growth regulators and growth factors as compared to the intact system. Therefore, any extrapolation from segment physiology to whole plant physiology must be considered with great caution. It is legitimate, of course, for a plant physiologist to analyze the growth regulators in the case of the isolated pea epicotyl or *Avena* coleoptile section; it is not legitimate, however, to assume as a matter of course that this undertaking will contribute to our knowledge about growth regulation in the intact system.

Just one further example to document that an analysis of the notion of causality is not only a matter of epistemology but also one that concerns the practicing scientist directly. Take cancer: cancer is determined by x factors in accordance with the scheme in Figure 5. Is it reasonable to state that smoking is the (or, a) cause of lung cancer? The ingredients of tobacco smoke are factors that can and will be decisive for the development of lung cancer if the other factors involved in the determination are available. There can hardly be any doubt that some constituents of tobacco smoke can act as factors in the development of cancer. However, it would be unwise even for practical reasons to envisage only these factors, while ignoring the other determinants.

I want to touch briefly upon the question, what are the logical prerequisites when we want to analyze the simultaneous action of *two* variables on a system (F_1 *and* F_2 in the general model). If the two factors interact with each other or with the rest of the system [(x − 2) factors] our analysis can only be descriptive and specific for the particular system and will probably not lead to any law-like general proposition. If, however, we can prove that there is no interaction involved, the system is open for a further logically transparent analysis, e.g., a molecular analysis. The first problem to be solved is therefore to prove that no interaction between the two variables comes into play. To make explicitly clear what we are doing we refer to the theoretical model in Figure 9, which states that if two factors (F_1, F_2) act independently on a system A, the response of which ("effect") is designated by $\Delta A'$,[3] there are only two possibilities for "no interaction":

1. If the two factors act independently on the same causal sequence, "multiplicative calculation" of the two factors will be observed. As a formula, $\Delta A'_{F_1F'_2} = a \cdot \Delta A'_{F_1}$

Expressed in words: the response (steady state rate, "effect") with two factors (applied simultaneously) is a *defined fraction* of the rate obtained with one factor irrespective of the magnitude of the rate $\Delta A'_{F_1}$. If *a* is > 1, the two factors act in the same direction; if *a* < 1, the two factors act against each other.

2. The theoretical alternative is that the two factors act on different causal sequences that lead fully independently to changes of A. An example would be that ascorbic acid is produced independently in two compartments of the cell and under the regulatory control of independent factors. In terms of the model (see Fig. 9): the factor F_1 controls the rate in compartment_1, the factor F''_2 controls the rate of ascorbic acid accumulation in compartment_2.

Under these circumstances, "numerically additive calculation" of the two factors will be observed. As a formula:

$$\Delta A'_{F_1F''_2} = \Delta A'_{F_1} \pm \Delta A'_{F''_2}$$

Expressed in words: a change of the rate of A, $\Delta A'$, which is caused by the two factors F_1 and F_2 (if applied simultaneously), is identical with the sum of the changes caused by the two factors if applied separately.

---

[3]The simple analysis we have in mind is valid only if $\Delta A'$ designates the change of some steady state rate. In the following example this boundary condition is met.

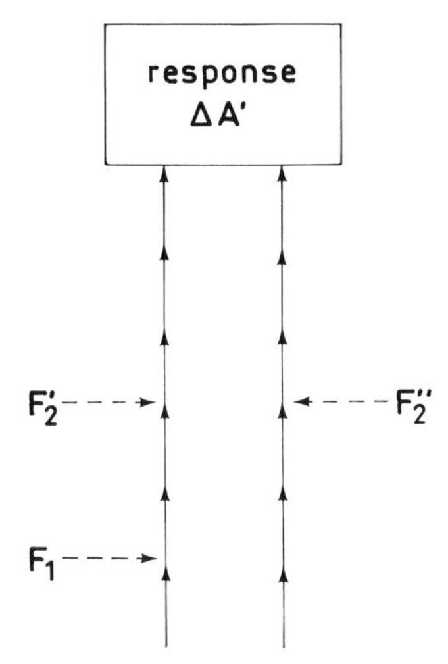

Figure 9. Model reflects situation in which two factors (F_1, F_2) simultaneously act on system A. Only those changes of system, responses $\Delta A'$, which can be conceived of as changes of steady state rates should be used in analysis. It is assumed that one and the same $\Delta A'$ can be elicited through two totally separate reaction sequences. Model is explained in text [modified after (5)]

In conclusion, multiplicative as well as numerically additive behavior of the two factors indicates "no interaction." Any other experimental outcome means "interaction," as a rule peculiar and system specific.

The following is an example of the multiplicative calculation of two factors. The system used is the anthocyanin-producing anabolic sequence in the competent cells of the mustard seedling *(Sinapis alba)*, the two factors are "light"[4] and the antibiotic chloramphenicol (CAP). Although CAP is a potent inhibitor of protein synthesis in bacteria it usually has little effect on protein synthesis in nonbacterial systems. Higher plants constitute an outstanding exception insofar as CAP depresses the synthesis of chloroplast protein, the synthesis of cytoplasmic protein being relatively unaffected. CAP within a certain range of concentrations (20–40 μg/ml) increases the rate of light-mediated anthocyanin accumulation in the mustard seedling. The initial lag-phase after the onset of light and the time of termination of anthocyanin synthesis are not influenced by the presence of the antibiotic (Fig. 10). With and without CAP a constant rate of anthocyanin synthesis is observed over a period of at least 24 h after the initial lag-phase. This is true for all light intensities (irradiances) investigated. It was found—and this is the decisive point in our context—that the

[4]The light factor is considered to operate exclusively through the effector molecule P_{fr} (active phytochrome). Since for the principal treatment this circumstance is without relevance, we use the term "light factor" and ignore the details of the mechanism of the light effect.

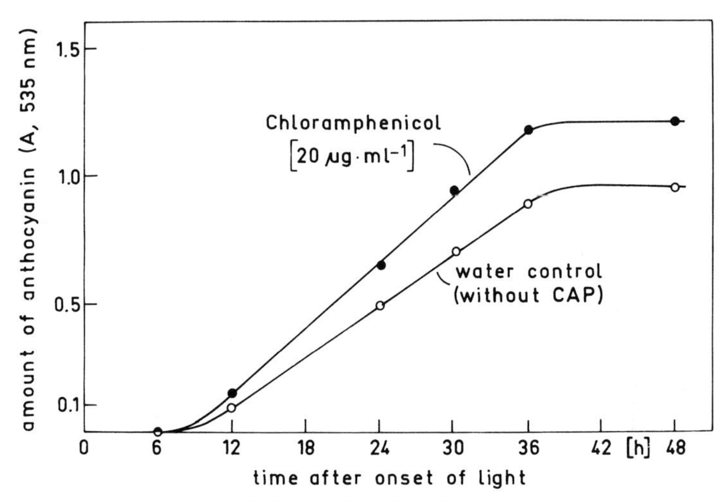

Figure 10. Kinetics of light-mediated anthocyanin accumulation in mustard seedling with and without CAP [after (6)]

relative increase of the rate of anthocyanin accumulation that is due to CAP (20 μg/ml) is independent of the light intensity (7). As a formula:

$$[\text{rate of anthocyanin synthesis}]_{\text{light, CAP}} = 1.52 \times [\text{rate of anthocyanin synthesis}]_{\text{light}}$$

This is true for a wide range of light intensities and thus rates of anthocyanin synthesis. It was concluded that light and CAP act as two independent factors whereby the action of light is indispensable, since CAP in total darkness does not cause any anthocyanin synthesis. The statements that light and CAP act independently on anthocyanin synthesis and that no significant interaction between the light factor and the remaining set of factors must be considered, opened the way to a thorough molecular analysis of the formula [amount of anthocyanin] = f $[P_{fr}]$ (see Fig. 8).

Returning briefly to Figure 8, we ask the question what this function means in terms of prediction. We consider the function

$$\Delta A = c \cdot [P_{fr}]$$

as an "empirical law" or at least as a law-like general proposition, valid probably for many systems in which phytochrome-mediated anthocyanin synthesis occurs. ΔA is the amount of anthocyanin "produced" by the amount $[P_{fr}]$ of active phytochrome; c is a system specific constant representing the boundary conditions.

If the boundary conditions (and therewith c) can be kept constant in a strict sense (this is possible, if appropriate seed material and technical facilities are available), we can predict with high accuracy the amount of anthocyanin that will be formed within 24 h if a certain amount of active phytochrome [P_{fr}] has been established by a brief light pulse at time zero. In order to make a precise prediction there is no need to know all factors involved in anthocyanin synthesis. The decisive point is only that all factors besides P_{fr} can be kept constant and can thus be considered as part of the boundary conditions. As already mentioned, the other factors besides P_{fr} may only be treated as boundary conditions if they are nonlimiting with respect to anthocyanin synthesis and do not interact with P_{fr}.

According to a classification advanced by KOCHANSKI (8) our system would have to be classified as a quasi-closed system. This term designates (in KOCHANSKY'S formulation) "a system in which the predicted event, occurring at time t depends fully and only on the initial . . . conditions which include both the state of the system and the border conditions as described at time t_0 at which the forecast is actually made." In simpler terms—if we know the factor dose generated at time zero by a brief light pulse (equivalent to the initial conditions) we can predict precisely the event occurring at time t if the boundary conditions are constant to the extent that the constant c in the foregoing equation really remains a constant. Under these conditions the forecast is very precise and not different, in principle, from a forecast in the classical quasi-closed-system situations in celestial mechanics or free fall, which serve as paradigms. KOCHANSKY'S second class, the "open system" (with respect to forecast) designates a system in which the occurrence of the predicted event depends not only on the initial state of the system (including the boundary conditions as described for time zero) but also on the subsequent particular interaction of the system with some external new environmental agent. It seems that philosophers believe that the open-system situation in forecasting is typical in most fields of biological inquiry whereas the quasi-closed-system situation in forecasting is typical for most fields of physics [8]. At present the philosophers are probably right.

7th Lecture

The Structure of Teleological Action

This is an old and venerable problem of philosophy (1). It is sometimes stated that a discussion of teleological action, i.e., goal-aimed, conscious, purposeful, and deliberate action, is outside the realm of science. This objection is *not* justified. Since man is a biological species, discussion of goal-aimed, *conscious* action and free will is in fact a legitimate topic of biology. Moreover, it is probable that some kind of teleological action occurs in the animal kingdom as well, at least with some higher mammals.

Another decisive factor in dealing with the structure of teleological action in our context is: when we consider the structure of science from the point of view of a philosophy of science, the structure of teleological action becomes a crucial question since the scientific enterprise must be considered in some way or other to be a particularly lucid and successful teleological activity. The scheme for teleological action I am going to describe may indeed be regarded as a model of the scientific enterprise.

The notion of teleological action designates a cardinal, venerable, philosophical, and existential problem insofar as teleological action implies free will and self-determined actions in the sense that we can create *de novo* determinants for our future actions and thus break causal continuity. Therefore, the question arises (it is one of the most important questions in philosophy *and* in the sciences) whether free will (in the foregoing sense) is compatible with our scientific attitude, which implies that a breach in causal continuity is not acceptable. From the point of view of science the *reality* of free will cannot be conceded. On the other hand, we definitely have the feeling that there is some freedom of choice concerning our actions and that the choices (or decisions) we make do have a real effect on our actions. Even a person who holds the opinion that the human mind is nothing else than the inevitable *correlation* of a highly complex physical system, the brain, and that free will is an illusion since our actions are fully determined by physical causality, will admit that the *feeling* itself, the notion of the free will, is a *real* phenomenon that cannot be disputed.

Teleological action implies not only consciousness, rationality, and free will; it also implies responsibility. Every human society considers the individual as being responsible for the choices he makes or the objectives he envisages, irrespective of the particular set of values and irrespective of the particular notion of "good" and "bad." At least in the case of a sane person, it is implied that the choices are made deliberately and in view of conscious anticipation of coming events and constellations.

The structure of goal-aimed, conscious, and purposeful action is illustrated in Figure 11. We consciously and deliberately anticipate a goal. This implies free will as well as the guidance by values and by our central propensity structure. We shall consider first the elements implied in goal-setting and then the integrated model.

We take the *existence* of values for granted at the moment, although at least some of these values might be irreducible spiritual values not explainable by scientific means. I acknowledge, of course, that modern ethology (2) and in particular the evolutionary concept of "inclusive fitness" [11] permit an explanation of values such as "altruism," "friendship," or "respect for elders" (3). However, it could be that there are at least some values that cannot be explained in full by any scientific theory. At the moment at least, some of the values of our culture must still be regarded as extrinsic values from the point of view of science and scientific theories. I emphasize, however, that only science can furnish a framework on which the extrinsic, spiritual values (if they exist) can exert their power. I will return to this aspect in the Lecture on Science and Values.

Human behavior is characterized by a central propensity structure that seems to be largely inherited. We think, argue, plan, and act along certain lines for *endogenous* reasons (see Lecture on Epistemology and Evolution). The value system and the central propensity structure are

Figure 11. Model illustrating structure of teleologic thought and action

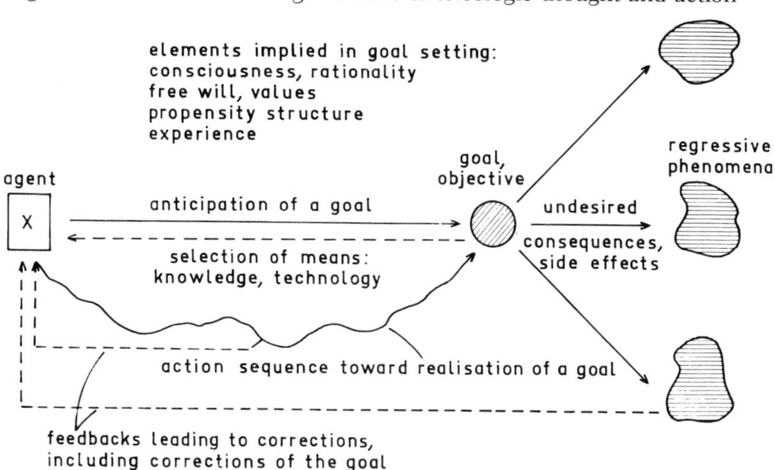

strongly interrelated, but in our context it seems warranted to consider them separately.

It is part of our central propensity structure (4) to think and to act in *expansive* systems. This is part of our genetic inheritance. It must have been a strong selective advantage for a hominid population in the course of natural genetic evolution to acquire the drive to grow, to expand, to migrate. Without this strong element in their central propensity structure, the hominid populations would hardly have walked from East Africa to all parts of the world. The strong factor in our propensity structure favoring growth and expansion is one part of the difficulties and obstacles we face in our efforts to establish steady state conditions in present human culture. The notions of population growth, economic growth, increase of gross national product, increase of national strength, etc., can hardly be replaced in the real world by ideas such as "optimization under steady state conditions" or "increase in quality" (at the cost of human labor and effort) instead of increase in quantity (at the cost of energy and nonrenewable resources) even though we *know* about the limitations to growth and the finite size of our planet. It has always been a horrible task to suppress inherited features of our propensity structure by intellectual persuasion. Under all circumstances, our inherited central propensity structure must be considered a strong element in all teleological activity. At least some elements of our propensity structure must be tamed to an extent that they do not render impossible reasonable teleological action in accordance with the present state of our civilization. Aggression, racial hate, ideological fanatacism are such elements of our central propensity structure that can no longer be tolerated. Aggression (the drive to conquer and even to exterminate other groups or populations), our inability to think *sincerely* in terms of mankind, our extreme preference for the group kept together by kinship, common history or political ideology *must* eventually be modified and redirected by strong intellectual concepts; otherwise our civilization will have no future. The threat of extinction is world-wide, but so far we have not been able to think seriously in terms of mankind or of global strategies.

Fortunately, it is not only our inherited propensity structure that tends to direct our teleological actions. Experience, in the sense of adaptive behavioral learning during ontogeny, including cultural tradition and social imitation, plays an equally important role in teleological action. However, learning is not independent of the central propensity structure. Learning, too, is a result of genetic evolution. As WILSON (3) puts it: "How can learning evolve? . . . What evolves is the directedness of learning—the relative ease with which certain associations are made and acts are learned, and others bypassed even in the face of strong reinforcement." The comparative study of learning has shown that various species have evolved different behavioral solutions. To some extent adaptive behavioral learning must be regarded as species-specific and thus as determined by *specific* genetic determinants above the level of a general learning ability.

Consciousness offers another problem. We may "solve" it by the hypothesis that consciousness is nothing more than a correlation of an exceedingly complex system such as is found in the central nervous system of man and in at least some other higher mammals. This hypothesis implies that any system as complex as the human brain or the brain of the higher mammals will *necessarily* show the characteristics of consciousness. This is an attractive hypothesis that seems compatible with all data available to *science*. If this hypothesis is considered "true," we must infer that *plants* will not develop signs of consciousness since their structural complexity (with regard to intercellular connections) must be regarded as being very modest, at least in comparison with vertebrates. While I agree, of course, that plants do not possess anything comparable to a central nervous system, I will later show in detail that plants do perform much better than is generally assumed with regard to perception and rapid processing of signals from the environment.

Returning to my central argument, I will now consider the structure of those human actions that are intentionally and consciously directed toward the obtainment of a goal. The model of teleologic thought and action in man (Fig. 11) only makes explicit a pattern with which we are all intuitively familiar. For example, many American proverbs show how familiar common sense is with the structure of teleological action: "Luck is often nothing but good planning properly executed." "When a known fact makes your plan unworkable, you are working on a bad plan." Teleological action is usually exemplified by the work of an artist. The artist aims at a creation that is adequate to express a concept formed in his mind.

The basis of teleological action is the conscious anticipation, the setting of a goal. The second step following anticipation will be the conscious selection of the means that seem to be required to reach the goal. Genuine knowledge and technology derived from it are the most important means at present, even in our personal lives.

The action sequence toward the realization of the goal is as a rule not straightforward. Continuous adjustments are usually required if the informational feedback tells the acting person or group of persons that deviations from the anticipated action sequence have occurred. Moreover, the realization of a goal will *inevitably* lead to undesirable consequences, called side effects. Since these side effects may grow into massive regressive phenomena, they must be thoroughly considered and compensated for as early as possible after they become detectable. This is an indispensable part of the responsibility of a person or a group acting toward the common good. Unintended, future dangers play the major role. It is the long-term effects that are the most problematic. Man is just not omniscent. As long as we exercise teleological activity, in other words, as long as we want to maintain human culture in some way or other, we must be prepared to take a risk, even the risk of self-extermination. The feedbacks indicated in the model will as a rule help us to minimize the risk, but it can never be completely excluded that gross errors in the estimate of future side effects and dangers will occur, in spite of concerned, responsible, competent anticipation, and consideration of undesirable regressive phenomena.

A major difficulty arises from the fact that in a pluralistic society there will be—as a rule—no consensus with regard to priorities and values. Teleological activity in a pluralistic society can only be based on a minimum common denominator. The different groups within a pluralistic society, the different nations within mankind, will differ at least with regard to the goals and values but as a rule they will also differ as to the assessment and evaluation of side effects or of potential or real regressive phenomena. There is no easy way to find consensus for teleologic action in a pluralistic society. The way to reach a reasonable compromise between different group interests is always tedious and sometimes very slow. We may dislike this state of affairs from our particular personal point of view, but it is superior to any kind of technocracy based on the dictatorship of a few people who believe that they possess the "right consciousness."

From the point of view of philosophy the most difficult problem in this model is posed by the elements free will and responsibility, since they imply a break in causality that cannot be conceded by the scientist. I want to take up this point once again.

As human beings, we depend on the belief that at least some of our actions *are* teleological actions, i.e., preceded by deliberation and choice, and that our choices can be influenced by considerations of consequences. Purely instinctive reaction instead of teleological action is not compatible with human dignity. For the sake of our self-esteem we must believe in freedom and free will. On the other hand, any break in causality cannot be accepted since such acceptance would overthrow science *and* the anticipation of goals. This has been one of the most serious dilemmas in the history of human thinking. It is neither a pseudoproblem [as MORITZ SCHLICK (5) tried to convince his contemporaries] nor is there a trivial solution. Let me pose the problem once again: we feel morally responsible for our willed actions and for our conduct *as if* we were able to create de novo determinants for our actions and thus give rise to some break in causal continuity. Most of us consider this feeling as an integral and indispensable part of human dignity. We know with subjective certainty that we possess free will. On the other hand, we know as scientists as well as from our daily life experience that the principle of causality cannot be questioned. There is no break in causal continuity. Science and teleological action itself depend on the unrestricted validity of the principle of causality. The problem free will vs. determination is a problem that cannot be solved by rational argument, by thought. (6). However, the theory of evolution permits at least some understanding of the fact that the rationally unsolvable problem, the dilemma, exists.

Man is a highly optimized and rigorously selected result of genetic evolution. This is not only true for his morphological, physiological, or psychological traits but also for those characters related to behavior, including group behavior and conduct within society. There are inherited traits of behavior in man (3,7).

Free will as a moral institution, i.e., free will as guided by moral instructions ("values," "duties," "commandments") must have been an extremely fruitful fiction in the course of genetic evolution since it facilitated the determination of the individual by rules of conduct, even by tough and uncomfortable ones, for the sake of a society or at least for the sake of the common good of a group of people. We feel free when we do what is necessary for the sake of the common welfare once we have become determinable by moral instructions. As Hegel puts it: freedom is the insight into necessity.

There must have been a strong positive selection pressure (with regard to "inclusive fitness," i.e., reproductive fitness of a group or a society) in favor of the fiction of free will. The notion of free will, which probably originated as a mutation within the central propensity structure, was thus anchored in the genes of man. This had led to the consequence that our genetic determinants force us to believe in free will although we *know* that there is no break of causal continuity in the real world. This is probably only a "modern," post-Darwinian version of Kant's (metaphysical) suggestion that man lives in two worlds and is subject to two different kinds of causation.

I would like you to consider briefly Figure 11 from the point of view of the structure of science. The scheme contains all elements we need in order to describe the scientific undertaking. In science, the goal of teleological activity is genuine knowledge, available as objective data and as theories. The means to reach the goal is logic and observation. The structure of inductive inference is fully compatible with this scheme if we consider a hypothesis (conjecture) as an anticipation of a theory, the actual goal. The feedback controls inherent in the scientific argument have already been described in connection with the "nature" of inductive inference (see Fig. 1). The intrinsic value system of science will be considered in detail in connection with the idea of a scientific community. The undesirable side effects and the regressive phenomena that might have to do with the scientific enterprise will be dealt with in forthcoming chapters on progress in science and science and technology.

As far as we know, plants do not exert teleological activity. However, the response of plants upon signals from the environment is sometimes so astonishing that some excited writers have granted attributes such as "mind" and "soul" or at least "emotions" even to plants (8). While this cannot be conceded by the experienced scientist (9), the perception and processing by plants of signals from the environment are startling enough. I will first describe the astonishing performance of a seedling in controlling enzyme synthesis under the influence of light and then return to the general question regarding the complexity and sophisticated status of the intercellular network in plants.

Figure 12 summarizes a major result of our studies on control of lipoxygenase synthesis by active phytochrome (P_{fr}) in the cotyledons of the mustard seedling *(Sinapis alba)* [10]. Lipoxygenase synthesis is controlled by active phytochrome (P_{fr}) through a threshold (all-or-none) mechanism. If the amount of P_{fr} exceeds the threshold level, lipoxygenase synthesis is fully and immediately arrested. If the amount of P_{fr} decreases below the threshold level, lipoxygenase synthesis is immediately resumed at full speed. This pattern of response is illustrated in Figure 13. The enzyme lipoxygenase increases in the dark (no P_{fr} present). Synthesis of lipoxygenase is suppressed as soon as a P_{fr}-level above the threshold is established by red light. A shift from red to far-red light at 1.5 h decreases the P_{fr} level below the threshold level (enzyme synthesis is resumed); a further shift from far-red to red light at 5 h increases the P_{fr} level above the threshold level with the result that lipoxygenase synthesis is again arrested, immediately and totally. It was concluded that in threshold regulation of enzyme synthesis the *primary* reaction of P_{fr} is characterized by a high degree of cooperativity. Pertinent models have been developed and supported experimentally. Threshold responses that imply a high degree of cooperativity in a matrix-ligand-interaction have previously been known only from nervous and sensory physiology where they play a major part. The threshold control by P_{fr} of the synthesis of an enzyme in a plant organ documents the fact that threshold responses can also occur in plants, at least in connection with the control of developmental processes.

Figure 12. Scheme describing concept of threshold regulation of lipoxygenase synthesis in mustard seedling cotyledons by P_{fr}. $[P_{tot}]_o$ is total phytochrome at time zero (36 h after sowing). This value is a constant. Amount of P_{fr}, $[P_{fr}]$, is always expressed as fraction of percentage of $[P_{tot}]_o$. Expressed in this way, threshold level, $[P_{fr}]_{th}$, is approximately 1.25% (0.0125) [after (10)]

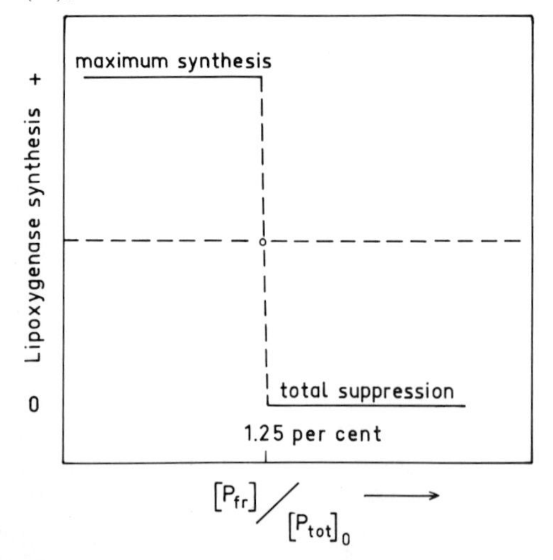

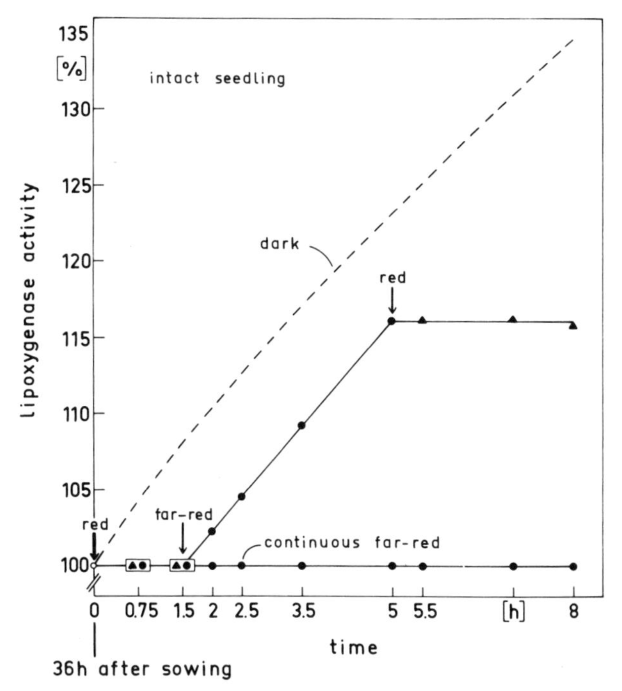

Figure 13. Time-courses of lipoxygenase levels in mustard seedling cotyledons in darkness (---), under continuous far-red light (●), and under irradiation sequence 1.5 h red light (▲)–3.5 h far-red light (●)–3 h red light (▲). While red light establishes P_{fr} level above threshold level under all circumstances, far-red light (given after 1.5 h red light) establishes P_{fr} level below threshold level. Far-red light *without* red light pretreatment establishes P_{fr} level *above* threshold level [after (10)]

A second fascinating feature of the lipoxygenase response is that the effector molecule P_{fr}, which exerts the threshold control over lipoxygenase synthesis, is located in the hypocotylar hook whereas lipoxygenase synthesis takes place in the cotyledons (Fig. 14). Thus a signal transfer between two neighboring organs is involved in this response. It was shown that the signal transfer from the hypocotylar hook to the cotyledons is rapid and precise and that the signal cannot be stored in the cotyledons. The signal transfer cannot be accounted for by hormones. Some biophysical signal must be involved. We believe that the morphological basis for the rapid and extremely accurate signal transfer is the intercellular connections known as plasmodesmata (12).

The suppression of lipoxygenase synthesis is, to my knowledge, the most precise cooperative reaction so far observed in plants. The hypocotylar hook has amazing properties as a receptor *and* processing organ for light signals from the environment. However, as compared to the performance of a nervous system, the system involved in the lipoxygenase response still looks very modest.

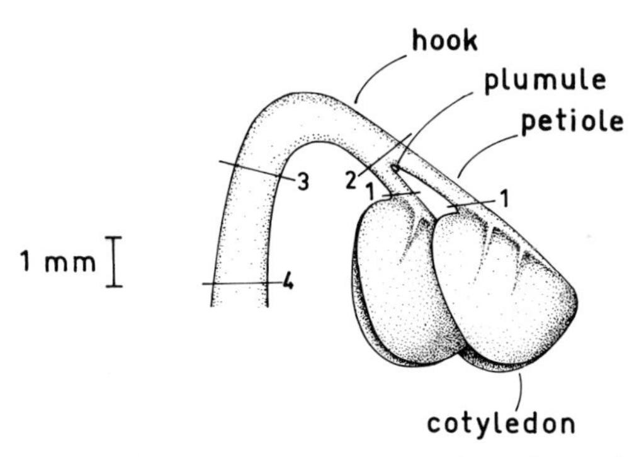

Figure 14. Upper parts of mustard seedling 36 h after sowing at 25° in darkness. With isolated cotyledons, no suppression by light of lipoxygenase synthesis can be detected. As soon as cotyledons are separated from hook, control of lipoxygenase synthesis by P_{fr} is completely lost. It was concluded that lipoxygenase synthesis in cotyledons is controlled by phytochrome located in hypocotylar hook [after (10)]

There is no reason to assume in plants a degree of system complexity and concomitant consciousness that would enable plants to act teleologically. Even the most elaborate reactions in plants must be described as "instinctive reactions." There is no reason (except zoological chauvinism) to reserve instinctive behavior for animals. Instincts are genetically determined intrinsic, species-specific patterns of reaction that can be released by specific key stimuli from the environment. If we accept the conviction that free will is fiction, it is doubtful whether "in reality" *we* are able to act teleologically. However, irrespective of our views with regard to the free will vs. determination problem, our highly sophisticated central nervous system, which permits the notions of consciousness, thought, and free will, has no detectable counterpart in plants.

8th Lecture

Physics and Biology: The Problem of Reduction

This topic is not only of deep-rooted theoretical interest; it also has a practical aspect. Research funds are dwindling rapidly throughout the world. The competition between the different fields of science will become stronger. Some fields will disappear completely. Departments will be closed. Under these circumstances, a fair comparison of the logical status of competing disciplines becomes crucial. If scientists are not willing to describe clearly the structure of and the relationship between the different fields, the future of science will be determined by prejudice and incidental experience of the members of legislatures, granting agencies, and governmental administrations. This implies a great risk.

The relationship between biology and physics (including chemistry) has been the subject of intense and sometimes even fierce debates over centuries. In principle, the question has been whether or not living systems can be explained in terms and laws of physics or, in other words, whether or not biology can be reduced to physics. The extreme positions have been called mechanism and vitalism. Mechanism (a very inappropriate term) was coined at a time when the view was held by some philosophers of nature that all science would ultimately be *reduced* to mechanics, i.e., to a particular branch of physics that considers matter in motion. This view has passed but the term mechanism has remained. A modern supporter of mechanism would postulate that the laws of present-day physics (in particular the laws of quantum physics) suffice to explain living systems if applied rightly. Vitalism implies that some nonmaterial elements of a spiritual nature (called "force of life," "élan vital," or—with reference to Aristotle—"entelechies") are constitutive for living systems but not for nonliving systems.

If vitalism means that a satisfactory explanation of living systems is beyond the capacity of science, then vitalism is fully obsolete. We do not know of any nonmaterial principle, restricted to living systems, that would principally forbid the reduction of biology to physics. Sometimes, however, the notion of vitalism has been used to emphasize the circumstance that some scientific laws might be required for the description of organized wholes in biology that cannot *principally* be replaced by physical laws. Since this view cannot be considered outdated, for the sake of clarity, the term vitalism should not be used to designate *this* position.

As far as I know the scene, many biologists who are familiar with the problem are partly reductionists and partly nonreductionists (1). This means that they favor a far-reaching reduction in some fields of biology whereas they maintain that a full-scale reduction of biology to physics is neither possible at present nor desirable for the future.

What does reduction actually mean? Reduction as a normative postulate of the logical empiricist is concerned with theories (2). A theory T_2 can or cannot be reduced to a theory T_1. A reduction of T_2 to T_1 presupposes at least that (1) the terms of T_2 can be *redefined* in terms of T_1 and (2) that all general propositions ("laws") of T_2 can be *deduced* from general propositions of T_1. Thus reduction is concerned with (theoretical) terms and laws (and thus implies a uniform scientific language, a problem by itself). Reduction is not concerned primarily with the observational level. On the observational level biology and physics will always remain different sciences. We have already mentioned that the relationship between theoretical and observational terms is defined by correspondence rules. We now add that the correspondence rules are specific and distinct for any field of science.

Does reduction actually work or is it a normative fiction of the philosophers of science? In practice, the *tendency* for reduction has had a successful record in the history of the sciences. Reduction satisfies one of the great desires of the scientist—to have unifying theories with a wide scope. A typical example is the reduction of classical thermodynamics to statistical mechanics or the creation of a few theories of great generality such as quantum mechanics or the theory of relativity. A considerable sector of traditional chemistry has been reduced to physics sensu stricto by the development of the electron theory of the chemical bond. These are impressive examples for successful reduction within the realm of physics. However, even here serious difficulties arise. The flourishing existence of chemistry as an autonomous science in the age of quantum theory clearly shows that it is still an impractical and inaccurate procedure to deduce the properties of complex molecules and reactions by means of quantum organic chemistry (14). It is a pragmatic postulate to grant chemistry at least the status of a semi-autonomous field within physics. This implies a direct measurement of molecular properties (observational level) as well as the use of these data for theoretical statements that are not *directly* related or immediately translatable into quantum theory. While some purists among the physicists will criticize this "semiquantitative" approach to the nature of molecules ("dirty chemistry"), the practical usefulness and success of the relatively naive "chemical" approach to the world of molecules cannot be denied. Thus the existence of chemistry as a semi-autonomous wing of physics is not in danger, although the claim of the physicist that quantum theory is capable *in principle* of supplying a quantitative foundation for the whole field of chemistry is fully justified. We conclude that a full-scale reduction of traditional chemistry to quantum theory is not seriously considered at present although it might be feasible in principle. The reason is that a reduction would be extremely difficult and impractical. Fortunately most scientists are pragmatic rather than dogmatic people. This pragmatism has been an essential feature in the difficult relationships between electromagnetism and mechanics. Faraday introduced the term (and the concept) of the electromagnetic field, and Maxwell developed the pertinent theory. As HEISENBERG (15) pointed out, " . . . the physicists began to understand that a field of force in space and time could be just as real as a position or a velocity of a mass, and that there was no point in considering it as a property of some unseen material called ether." However, the traditional belief among physicists that reduction to the mechanical concepts could finally be effected was a strong hindrance. As HEISENBERG (15) continues, "it was not before the discovery of relativity that the idea of the ether was really given up, and thereby the hope of reducing electromagnetism to mechanics."

Some critics of the normative reduction scheme introduced above [e.g., KUHN and FEYERABEND (3)] have emphasized that in practice a *replacement* of theories by other theories prevails rather than strict reduction. This is certainly also the experience of the practicing scientist. The critics of the normative reduction scheme even argue that in the history of science a strict reduction in the normative sense has never occurred. Indeed, in actual science new theories rarely displace their predecessors entirely. Often there is a smooth continuity between the old and the new theory, in particular if the old theory is a special case of the more general new theory.

As Einstein discovered in 1905, Newtonian mechanics fail when bodies move very fast relative to each other. In such cases, Newtonian mechanics has to be replaced by relativistic mechanics. However, Newtonian mechanics remains far superior for the class of phenomena for which it is valid. There can hardly be any doubt that relativistic mechanics does not equal Newtonian mechanics in mathematical perfection and elegance. While Newtonian mechanics has been degraded (" . . . no more than a low velocity approximation to the new theory of relativistic mechanics"), it will remain (probably forever) the theory of choice whenever we deal with low velocities.

Another more personal example: when I received my training in physics, classical thermodynamics and statistical thermodynamics were taught by the same professor in the same course with nearly the same weight.

In present-day biology, the coexistence (in general, a *peaceful* coexistence) of Mendelian and molecular genetics is quite impressive. There are several excellent textbooks that competently treat both aspects without trying to enforce reduction of Mendelian genetics to molecular genetics by all possible means. As a rule, the problem of a consequent and rigorous reduction does not seem to be an urgent one for most biologists. It is the philosopher of science, the logical empiricist, who stresses reductionism. In fact, since nobody assumes that entelechies play a part, in living systems, there is no a priori reason why the simultaneous use of molecular and purely biological constructs and concepts in genetics and developmental biology should not be permitted. However, in my opinion the terminological slovenliness often goes too far under these circumstances and represents a very real threat to the seriousness of the whole field of transmission and developmental genetics. For example, it has become difficult, sometimes even for the expert (not to mention the student), to find out what the definiens is for the term gene in a particular talk, discussion, or paper. This is mostly not due to a lack of logical resolving power on the part of the individual (although this may contribute sometimes to the terminological confusion), but to plain carelessness and even snobism. This is irresponsible. A *precise* definition and use of terms is, of course, essential in a field of research in which different theories are merging and slowly and gradually replacing each other. Under these circumstances one may not expect a context-invariance of terms as a matter of course.

Another negative consequence of the happy coexistence of competing theories (in negative terms, of the lax treatment of the actual problem) is that the essential points of the scientific matter become darkened. In the case of molecular biology vs. classical genetics, the essential points seem to be the following (Fig. 15).

Figure 15. *Upper part:* Some examples for successful reduction within physics. *Lower part:* Three steps in process of reduction of classical (Mendelian) genetics to quantum physics

T_2:	theory to be reduced	classical thermodynamics	Newtonian mechanics
	↓	↓	↓
T_1:	reducing theory	statistical mechanics	relativistic mechanics
T_2:	classical genetics	molecular genetics	macromolecular chemistry
	↓	↓	↓
	molecular genetics	macromolecular chemistry	quantum theory

classical genetics
↓
quantum physics

1. Can Mendelian (classical) genetics be reduced to molecular genetics?—It is sometimes suggested that this reduction would be possible as a matter of course if really desired. This suggestion is at least premature.

2. Can molecular genetics be reduced to macromolecular chemistry?

3. Can macromolecular chemistry be reduced to physics (i.e., to quantum theory)?

I think that strict reduction [in NAGEL'S sense (3)] is not feasible in any of the three cases, at least at present.

Let me comment briefly on the different aspects:

3. The exact treatment of complicated aperiodic macromolecules with quantum mechanics still causes great difficulties. While these difficulties are not insurmountable in principle, in practice the theoretical treatment still depends on approximating methods that are not completely satisfactory and difficult to handle. We have considered this aspect in a previous section about the relationship between physics sensu stricto and chemistry.

1. While nobody doubts that any Mendelian gene has a molecular equivalent (nominal definition: a structural gene is a segment of a DNA macromolecule that can be transcribed and that codes for a polypeptide), the sincere (i.e., nonsuperficial) deduction of Mendel's laws from molecular genetics poses serious and apparently insurmountable difficulties. One of these is that the two theories still use very different terms (even though they might use the same word, see above) that cannot easily be reduced to each other. For example, in molecular genetics the gene is no longer the unit of mutation and cross-over. Rather it is the nucleotide (or a few consecutive nucleotides) that must be regarded as the unit of mutation or cross-over. Even in purely biological (i.e., nonmolecular) fine structure genetics, the unit of function is the cistron, which differs from the unit of mutation (muton) and the smallest unit of cross-over (recon). As criticized previously, many biologists tend to be very flexible (a more appropriate term would be lax or careless) in this area, talking about genes even though they mean DNA segments or cistrons.

On the level of laws or law-like structures, the full-scale reduction of Mendelian genetics to molecular genetics would be a hopeless undertaking at present. An example will indicate the crucial point: it is stated in nearly every text that Mendel's laws depend on the "random" distribution of the chromosomes during meiosis. If a diploid mother cell possesses 8 × 2 chromosomes, each of the four meiospores or gametes will have 4 chromosomes. This so-called "random" distribution of the chromosomes is, of course, not a statistical (or probabilistic) phenomenon but subject to a strict law, which can be formulated as follows: if one of the homologous chromosomes moves in one direction (toward one pole), the other one *must* move in the opposite direction (if no "error" or disturbance is involved). It is the "operation" of this law that guarantees an equal distribution of the chromosomes over meiospores or gametes. If every chromosome of the mother cell would have a 50% chance to move to one or the other pole, irrespective of the movement of its homologous partner, there would be hardly any chance for an equal distribution of the chromosomes over the products of meiosis. The only random event involved in meiosis is the decision as to which one of the two homologous partner chromosomes moves first toward a pole. If this decision has occurred, the partner no longer has any choice. In this way an apparently random distribution of chromosomes and genes can occur during meiosis even though the actual number of chromosomes involved, say 2n = 8, would be far too small to achieve a random distribution by purely statistical means. In other words: a strict biological law is the basis of the "random" distribution of genes during meiosis. This empirical law is also obviously an antecedent condition for population genetics, since it is the essential prerequisite for random behavior in recombination. There is not the slightest chance at present to reduce this law to any "law" of molecular genetics.

Another argument against reduction of classical genetics to molecular biology is that classical genetics is the only part of biology that has been successfully axiomatized [by WOODGER (4)] using the formal apparatus of WHITEHEAD and RUSSELL'S *Principia Mathematica*. A similarily strict axiomatization has not been performed so far in the case of molecular genetics. With regard to *axiomatization* classical genetics are clearly superior to molecular genetics.

On the other hand, the scope of classical genetics has been much more narrow than the range of molecular biology. Classical genetics has been primarily transmission genetics; molecular biology intends to cover transmission genetics, including identical replication of genes, as well as developmental genetics. Developmental genetics is concerned with the causation of traits by genes during development. In this sense, classical genetics could indeed be regarded as a special, very limited case of the more general new theory. Any adequate gene model in molecular biology must satisfy the requirements of transmission as well as of developmental genetics. As you all know, the Watson–Crick model of genetic DNA from 1953 (5) serves both functions. It not only explains the replication of genes but it is also consistent with the various models of gene function or gene expression, e.g., with the Jacob–Monod model from 1961 (6), which explains adaptive enzyme formation in bacteria. In the higher eukaryotes, which possess chromosomes and effector molecules such as hormones and phytochrome, the models to explain differential gene expression are much more complicated but nevertheless consistent with the Watson–Crick model. All models for gene expression are based on the one-gene–one-enzyme hypothesis (today, one-cistron–one-polypeptide hypothesis) introduced originally by BEADLE and TATUM in the forties and on the universal validity of the so-called genetic code, a set of rules of correspondence that determine the correct translation of a peculiar linear nucleotide sequence into a peculiar linear amino acid sequence.

These few statements may suffice to indicate the tremendous progress molecular biology has brought to biology. With regard to the problem of "reduction," I feel that molecular biology will *replace* in practice classical transmission genetics as far as possible without any enforcement. Classical genetics will not be fully displaced since for many statements in transmission genetics the terms and laws of classical genetics are more useful than the corresponding statements of molecular biology. We remember that the predictive power of classical genetics is very great, although this section of science is exclusively based on strictly biological constructs.

2. Can molecular genetics be reduced to macromolecular chemistry? This is obviously the *crucial* question. Since molecular genetics (which always includes developmental genetics), presupposes the function of the cell, the question can be reformulated in more general terms: can the function of a cell or of an organism be reduced to chemistry? The answer is, no. The reason is that every living system is characterized by "organized complexity" whereby "organized" denotes that the components (elements, parts) of the system *do not* form a homogenous, randomized aggregate. Rather, any living system is a highly improbable system that contains a large amount of "organizational information" (or, neg-entropy, spatial information, system information). Much of this specific, peculiar system information will be irreversibly lost if the system is destroyed. For example, if a cell is homogenized much of the system information is gone (some will be retained in the more or less intact organelles such as mitochondria or plastids). "They all look the same in the Waring blender" is an old saying of the (innocent) reductionists among the biochemists. This is precisely true: a yeast cell, a frog cell, or a human cell all look very much alike after homogenization. The crucial point, however, is that this biochemical advantage (e.g., for the study of basic cell components and fundamental cell functions) is paid for by the loss of essential information about the *organization* of the living system. In brief: analysis of isolated elements inevitably results in the loss of information about the particular and highly nonrandom *relationships* between the elements (parts) of the system.

An example: squeeze an orange, take the juice, and add it back to the residue. You will not detect the slightest tendency for a reconstitution of the orange even though not a single physical element of the original orange was lost. Why? The reason is that during squeezing the particular and unique system information (organizational information) was destroyed. Without this information there is no possibility of a reconstitution of the original status of the system even if free energy is supplied in appropriate form. The *unique* system or organizational information (you cannot mistake an orange for any other fruit) was introduced during the development of the fruit. Originally the organizational information was derived from *genetic* information. The kind of "self-assembly" from the parts, which works with the tobacco mosaic virus (16) and partly with the T_4-bacteriophage no longer works, for reasons of principle, on the cellular level. The construction of a protein molecule is a simple example of "directed assembly": the amino acid sequence is absolutely nonrandom. It is dictated by the nucleotide sequence of genetic DNA via messenger RNA. The question (repeated once in a while) whether or not "life" can be created de novo in the test tube out of a soup of molecules is out of date. We *know* that a living system will not appear in the test tube if the right physical elements (including appropriate sources of free energy) are just mixed. The limiting factor is *information*. It is believed that this information can be supplied in principle by DNA and by those regulatory factors that are required to obtain an ordered and sequential transcription and translation of the genetic information. In other words: the creation of a living system in the test tube implies that we are able to imitate the process of *development* in the test tube. The source of information and the control elements for the *ordered* use of the information during the development of the system must be supplied from outside. Both the source of information and the control elements must be obtained from living systems since there is no chance at present to synthesize these items by chemical means independently of any living system. So far the synthesis of genes (e.g., for hemoglobin or for the hormone angiotensin) cannot be performed with purely chemical means but only with the aid of enzymes obtained from living systems. At present, it is an open and purely speculative question to what extent and when we will be able to imitate "development" in the test tube. What we know for sure, however, is that the orderly construction of a living system depends not only on the availability of the physical elements (building blocks) and of free energy (power) but also on the availability of the pertinent information (blueprint). Moreover, the construction depends on those factors that can read the information and are able to guarantee an orderly use of the information (analogous to a fantastic master builder). At least in the foreseeable future the master builder must be obtained from preexisting living systems.

In summary, a living thing cannot be explained in terms of its parts but only in terms of the *organization* of those parts. Although the whole is nothing but the parts put together, the "putting together" creates something that cannot be accounted for by the properties of the parts themselves. The question is, where does that organization come from? The answer is: it stems from genetic information that has been invested during the construction of the organization during development. Every process occurring within living systems is (in principle) compatible with a physical mechanism. However, if we want to understand how the particular living system came into being we must refer to development guided by genetic information.

The *analogy* with teleological action, indicated in the foregoing section by the terms in parentheses, may be emphasized once more by considering the development of a car. A car is not more than the sum of the parts if (and this is the decisive point) the sum of the parts includes the spatial relationship (the "organization") of the parts. This organization has nothing to do with the parts themselves. It cannot be derived from the properties of the parts. It is specific and peculiar information that was introduced during the construction of the car. It can be described once the car is ready for use, as the system or organizational information, characteristic for this particular model of a car. In general terms: technological systems must always be considered as the result or product of teleological action (see Fig. 11). Therefore one may *not* assume that the specific structure of a machine is explainable by the laws of physics (7). The specific design is completely *extrinsic*. But, of course, every process occurring within the machine is compatible with a physical mechanism. In an *analogous* sense the specificity and the "design" of biological systems cannot be explained by the laws of physics. It is the specific and peculiar genetic information that determines the specificity of organismic organization. Only if evolution and ontogenetic development could be explained fully in terms and laws of physics, could an explanation sensu stricto of any real organism in terms and laws of physics be envisaged.

The circumstance that the destruction of a system is accompanied by the loss of system information that can no longer be recovered from the elements is, of course, *not* peculiar for biology. Instead of a crashed plane we chose a less macabre example [adopted from RUSE (2)]: Take a system with two resistors, the resistances of which are R_1 and R_2. It makes a great difference whether these two resistors are put in series (the law is then $R = R_1 + R_2$) or in parallel (the law is then

$$\frac{1}{R} = \frac{1}{R_1} + \frac{1}{R_2}$$

where R is the total resistance of the system). If the system is taken apart, and we investigate only the isolated resistors, the original system properties cannot be deduced from the elements. In other words: it cannot be decided on the basis of the isolated elements which one of the two preceding laws really describes the original state of the system.

The following example is taken from molecular biology or from macromolecular chemistry, depending on the point of view. The random nucleotide composition of a particular piece of DNA no longer contains the sequential (genetic) information characteristic for the native DNA molecule. Moreover the sequential information of a DNA molecule has a historical dimension since the particular DNA molecule was synthesized with the aid of an ancestor molecule, i.e., *not* in a random process. It is this unique historical dimension of the aperiodic macromolecules that so far resists reduction to the laws of macromolecular chemistry.

A new kind of reduction, which has attracted much interest recently, is to *simulate* in a computer program some properties of a particular biological system. Simulation is a general term applied to the process of conducting experiments on a model instead of attempting the experiments with the real system. In biology, simulation studies were made in particular with the use of computer simulation techniques. Computer simulation models are in a way intermediate between concrete empirical studies and the mathematical analysis of hypothetical, abstract models. In biology, simulation means setting up in a (digital) computer conditions that describe with reasonable approximation relevant properties of the biological system we want to study. On the basis of the input data (objective data as well as assumptions) the computer then generates the resulting information (output). Different input data and assumptions can thus be tested to determine their effects on the behavior of the system. As an example, the simulation of the process of glycolysis has already advanced quite far (8). While the predictive value of simulation can be high at times, it is generally felt that simulation should only be regarded as an *additional* tool. It cannot be a substitute for an explanation in terms of the actual constituents of the living system.

With regard to psychology, the use of mathematical models and computer simulation could lead to a breakthrough (9). It is generally accepted that neurophysiology is the material basis for psychology. However, a serious effort to reduce psychology to neurophysiology has not been made so far (18). This kind of enterprise would first of all require a redefinition of many psychological terms in terms of neurophysiology. Many psychologists are obviously reluctant to consider seriously this formidable task, in particular since the entry of psychoanalysis into the academic scene. At this point, computer simulations of some functions of the human brain might be helpful. Most natural scientists feel that the human mind is "nothing more" than a correlate of a very highly organized material system, the human brain. In CRICK'S (admittedly blunt) words: "I myself, like many scientists, believe that the soul is imaginary and that what we call our minds is simply a way of talking about the functions of our brain" (10). This assumption implies that a computer of gigantic system complexity approaching the material system properties of the human brain would show at least some of those properties that we usually attribute to the construct "mind".

Is the enforced reduction of as many biological fields as possible really desirable? I think it is not. I argue against excessive "reduction" even in the sense of replacement because I am afraid it would impoverish biology. The diversity of biological phenomena requires a diversity of scientific approaches. The physiologist's look at a system cannot be replaced by the comparative biologist's look *at the same system*. A causal explanation and a functional or teleological explanation are complementary rather than mutually exclusive.

Not only is there no reason to eliminate biology as an independent discipline in favor of physics; there is also no reason to reduce the diversity of scientific approaches within biology as long as these approaches are scientifically sound *and productive*. If, however, a field becomes dominated by rigid paradigms with a concomitant shortage of new ideas, a drive for reduction can stimulate a revolution! Classical and molecular genetics are good examples.

At this point of our discussion, I want to remind you of the aim of biology: the aim of biology is the explanation of living systems and their diversity. This implies an understanding of the living state. The living state is a property of matter in a *certain state of organization*. As long as the organization is preserved the living state will be preserved. Some striking experiments in this regard were done by BECQUEREL (11). He dessicated various species of *Rotatoria*, *Tardigrada*, algae, spores of bacteria and yeast, fragments of lichens and mosses, and seeds of various plants. Next he cooled them down to a temperature of approximately 0.01 K (this is 0.01° above absolute zero at −273.15°C) and held them in this state for 2 h. Then he brought them gradually back to normal conditions, and all the organisms revived. There is agreement that at temperatures very close to absolute zero no life *processes* go on, even slowly. There is no gas or fluid state, all elements become solid. The only explanation for BECQUEREL'S data is that the molecular organization within the cells and organisms did not irreversibly change during the period at nearly absolute zero, and thus the living state was preserved. The cells could function again normally after being brought back to normal conditions. There are many similar experiments, including those by DOMBROWSKY (12) who was able to bring to life again various kinds of bacteria that were enclosed for several hundred million years in salt obtained from Zechstein salt mines.

With regard to the "nature" of the living state, I conclude that every process occurring within living systems can (in principle) be explained by a physical mechanism. I further conclude that if we could reproduce experimentally EIGEN'S (17) stochastic explanation of primordial evolution of biological macromolecules and build up the specific molecular *organization* of some living state in the test tube, we would have made some kind of a living system. There can be no doubt that this objective is unrealistic at present; however, there is no reason to assume that we will meet any *principal* obstacles on our way to a man-made living system of some primitive kind. The question is whether this goal is really attractive once we are all convinced that it could be obtained. It might be more worthwhile to elaborate the laws and principles that govern the behavior of those living systems that have actually been created in huge diversity and utmost sophistication by natural evolution.

At least implicitly, the reductionist's attitude is heavily objected to by the radical critics of present society—such as Roszak in the USA or Habermas in Germany—who look at reduction as a new malice of the "mechanistic" Baconian approach to nature and consider science as a destructive source of domination and control over nature and man. This attitude is based on a deep misunderstanding of the nature of science, and we could ignore it if the dominant theme in the attack on science were not the accusation that science is a threat to human values. We will take up this point again in the Lecture on Science and Values.

At the moment, I only want to state that with regard to reduction the same argument we used previously for biology also applies to the science of man. An excessive biologism in the science of man would lead to the same impoverishment of this discipline as would an excessive physicalism in biology. However, the postulate that the science of man is subject to the same norms of the ethical code of science as physics or biology may not be violated. Only those statements that are truly *scientific* statements can be considered as relevant in the rational dispute about the nature and destiny of man. At this point I want to mention the principle of complementarity as developed by BOHR in 1927 (13). Certain observations about light require a particulate interpretation of light; other observations are only consistent with the notion of a wave-like "nature" of light. Neither a particulate nor a wave theory of light accounts for all observations about light. The principle of complementarity says that the two theories about light are not reducible to each other. Rather, they complement each other. Although we nowadays agree that the Schrödinger equation takes care of both, the particulate as well as the wave aspect of light, the fact remains that it depends on the experimental situation whether we consider light as a photon or as a wave.

Complementarity in biology has a different meaning. The term designates the fact that different approaches are necessary to fully describe a particular living system, e.g., biochemical analysis of the elements, physiological black-box experiments (input–output experiments), in vitro experiments with isolated organelles, microscopic analysis of the living system, light microscopic and/or electron-microscopic observations of the fixed and appropriately "stained" system. Information from all these levels is required and indispensable in order to obtain a balanced description of the system. This is the sense of complementarity in biology.

9th Lecture

Physiology and Comparative Biology

The field of biology can be divided into two areas: experimental biology and comparative biology (Fig. 16). The aim of experimental biology is causal explanations of biological events; that of comparative biology is functional or teleological explanations of biological structures, processes, and molecules. The two areas of biology are complementary rather than competitive or mutually exclusive.

Sometimes a third area is defined as descriptive biology. An important field in descriptive biology is that sector of biochemistry concerned with the description of the molecular constituents of living systems. Both taxonomy and experimental biology depend heavily on this field of research. Since as a rule, descriptive biology is *also* comparative it can be regarded as belonging to the wider realm of comparative biology.

At present, experimental biology consists of two strongly competitive fields, biochemistry and physiology (Fig. 16). In biochemistry the tendency is analytical. The field tends toward in vitro measurements and in vitro experiments. The complexity of systems that can be studied biochemically is limited. Organelles, mostly damaged during isolation, are the most complicated systems that can be handled by biochemistry at present. A biochemical treatment of the construct "living cell" is not feasible. This matter has been discussed in the previous lecture.

In physiology the tendency of the researcher is directed toward a conservation of a largely undisturbed system complexity. In vivo experiments are preferred, mostly performed as black-box experiments in which an output of the system is measured as a function of some input while the boundary conditions are strictly defined and controlled. The behavior of the real system is usually described in terms of *formalized* systems, sometimes in analogy to technological systems, as in the field of cybernetics.

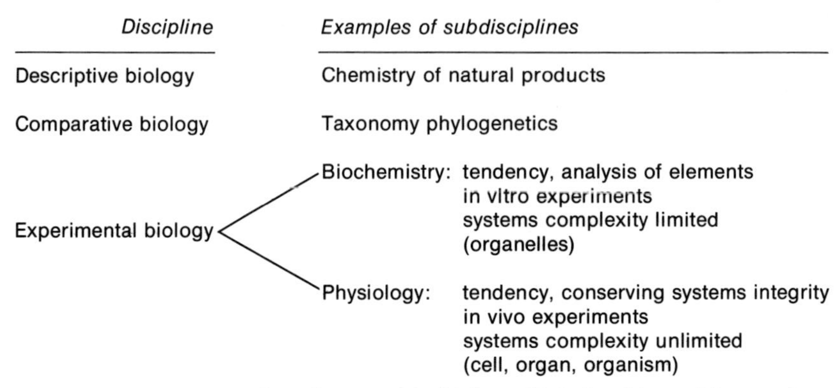

Figure 16. Major disciplines within biology. Relationship and interaction between different disciplines is discussed in text [after (1)]

Some fields of physiology prefer to define the output and input of the systems they are studying in chemical or physico-chemical terms. In those fields physiologists often use biochemical, in particular analytical methods, and thus are dependent on the progress of analytical biochemistry. Thus, the interaction between the different disciplines is much greater than Figure 16 suggests at first glance.

In any case physiology is a *system* science. The application of the systems concept has been essential for this science from the beginning. We have already mentioned that the universal law $\Delta G \neq 0$ is valid for all "open systems." Since all living systems are open systems, the principle is valid for all living systems. The application of irreversible thermodynamics to open systems [in particular by PRIGOGINE (2)] has led to a number of general propositions that refine the somewhat trivial statement $\Delta G \neq 0$. A general proposition in the context of Prigogine's theory states that in any open system (and therefore in any *living* system) the increase of entropy per unit time is minimum if the system is in steady state. Another formulation of this principle is that the rate of energy devaluation in an open system decreases if the system approaches steady state and is smallest in the steady state. It is inferred from this principle that steady state situations are favorable for a living system with regard to conservation of useful energy. It is further inferred that there must have been a selection pressure in favor of steady states in evolution. This is an example of a *causal* explanation of the existence of preferred states in living systems. The Prigogine principle explains without any teleological implication ("teleological" in the sense of *external* teleology) why a certain preferred state or goal-state is attained in the face of a wide range of *possible* states. On the other hand, the process of *development* (which includes differentiation and morphogenesis) must be considered as a constitutive *deviation* from steady state. It is inferred that the rate of energy devaluation will increase under these conditions.

Measurements of the energy yield during photomorphogenesis are consistent with these predictions from PRIGOGINE'S principle (3).

This may be considered only as a hint indicating the usefulness of the systems approach in physiology. I now want to describe in some detail the systems concept in plant physiology by describing some properties of the phytochrome system (4).

The normal development of a higher plant takes place only in the presence of light. A plant grown in the dark shows the characteristics of etiolation. Normal development in the presence of the light factor is called photomorphogenesis (3). The phenomenon of photomorphogenesis has been studied intensely by many plant physiologists. I refer particularly to photomorphogenesis of the mustard (*Sinapis alba* L.) seedling, the favorite subject of research in our laboratory (5).

The environmental factor light controls development of the mustard plant through the phytochrome system (Fig. 17). The phytochrome system as it occurs in the cotyledons and in the hypocotylar hook of the mustard seedling can be described quantitatively by a model that consists of four elements (components) and four reaction constants. In the dark-grown seedling only P_r will be found. This element originates from a precursor P'_r via a zero order reaction. This means that there is a constant rate of P_r formation throughout the period of our investigation. P_{fr}, the physiologically active form of the system (the *effector* element) appears only in the light. It originates from P_r. The photo conversion

$$P_r \xrightarrow{^1k_1} P_{fr}$$

is photo reversible. In both directions the photochemical reactions are of the first order. P_r is quite stable in the dark, P_{fr} is not. The irreversible destruction of P_{fr} to P'_{fr} is also a first order reaction. The half-life of P_{fr} in the mustard seedling is 45 min (boundary conditions: 36 h after sowing at

Figure 17. Model of phytochrome system as it occurs in mustard seedling cotyledons and hypocotylar hook

$P'_r \xrightarrow{^0k_s} P_r$ represents de novo synthesis;

$P_r \underset{^1k_2}{\overset{^1k_1}{\rightleftarrows}} P_{fr}$ represents the light reactions;

$P_{fr} \xrightarrow{^1k_d} P'_{fr}$ represents the dark destruction

[after (11)]

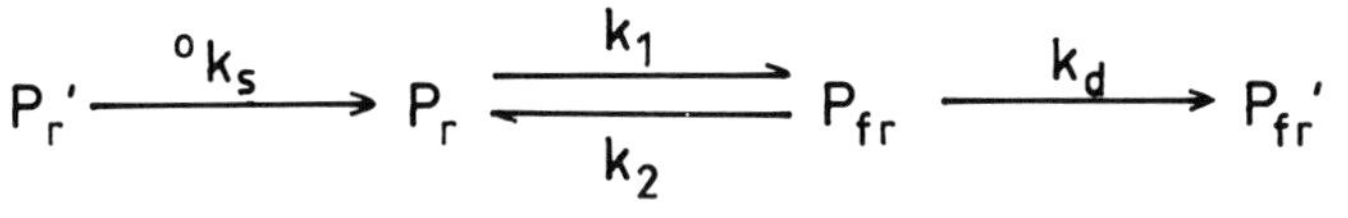

$$P_r' \xrightarrow{^0k_s} P_r \underset{k_2}{\overset{k_1}{\rightleftarrows}} P_{fr} \xrightarrow{k_d} P_{fr}'$$

25°C; germination and growth of the seedling under strictly standardized conditions).[5]

The phytochrome model is pertinent for illustrating some points of view that are relevant for modern physiology.

A system in physiology is an intellectual creation from elements (components) that are related to each other by defined interactions. In vivo experiments must prove that the model (the intellectual creation) represents the real system with satisfactory approximation. If all elements of a system and the manner and quantity of relationships between them are defined, one can derive all possible properties of the system. This kind of deduction is called systems analysis.

In the present case the quantitative relationship between the elements is described by the reaction orders and the magnitude of the reaction constants. Two of the reaction constants depend on light, i.e., their magnitude is a function of wavelength and irradiance of the incident light (${}^{1}k_{1}$ and ${}^{1}k_{2}$). The reaction constants ${}^{0}k_{s}$ and ${}^{1}k_{d}$ are independent of light.

The "interaction" between the elements leads to the consequence that the system exhibits properties that can in principle not be recognized, if we look at the isolated elements only. These system properties are thus the manifestation of "organized complexity" (see previous lecture). It is immediately obvious that even the most detailed study of the elements in vitro will not yield any knowledge about the system properties. A rigorous analytical approach that destroys the organized complexity will lead to the loss of precisely those properties that the physiologist wants to know. It is obvious that a successful molecularization of the elements is not sufficient (in the case of phytochrome we are dealing with a photochromic chromoprotein); only the complementary investigation of the system properties in vivo makes the molecularization of the elements physiologically relevant.

---

[5]The term "model" has been used so widely that it lacks a single precise meaning. The term covers the abstract "model" of the theoretical physicist as well as the mechanical "model" of the engineer, e.g., a "model" airplane. Intellectual inventions such as point masses, perfect conductors, ideal gases, have also been designated as models. Analogy "models" are supposed to embody features considered essential to the system we want to describe. The venerable chromosome theory of inheritance offers a particularly interesting case of an analogy model. A chromosome chart represents the genes as lying in a line—like beads on a string, as Morgan put it. This is an analogy model. Of course, genes are not really beads on a string, but genes were considered to be analogous to beads on strings in some essential features, namely, their discreteness and colinearity. In physiology, the specification of the elements of a physiological model in terms of physicochemistry has been a major concern. An outstanding example has been the inquiry as to the "chemical nature" of the gene construct. The objective has been to obtain a "molecular model" of the gene. The Watson-Crick model of the genetic DNA (the double helix) considered not only the chemical knowledge about DNA, but also the data obtained from X-ray crystallography of DNA crystals. The way Watson and Crick put together the model (29) must be considered as the construction of a relatively crude analogy model, put together from mechanical parts. However, since the model could be treated with some approximation in terms of quantum physics, it rapidly reached the status of a formal model similar to other molecular models in physical chemistry or quantum organic chemistry.

This is a general proposition! One could demonstrate this point with every well-known physiological system, including the cytochromes and the respiratory chain, chlorophyll and the electron transport chain of photosynthesis, oestrogens and the model of the hormonal control of the menstrual cycle.

The phytochrome model shows some further features that merit consideration: (1) It *cannot* be understood as a homeostatic feedback model; as an example, the reaction that leads to P_r is of zero order under all circumstances. This implies that it is neither regulated from P_r nor from P_{fr}. Obviously the phytochrome system escapes a description in terms of the usual negative feedback models of the biological cyberneticist. (2) The phytochrome system is an excellent example for a steady state system. The properties of the steady state can easily be derived by very simple systems analysis.

The phytochrome system is in a steady state if the rate of de novo synthesis of P_r is equal to the rate of P_{fr} destruction. Under these circumstances $[P_{total}]$, i.e., the sum of the amounts of P_r and P_{fr}, is constant. It is the equation $[P_{total}]$ = constant that defines the steady state of the phytochrome system. In symbolic language the foregoing verbal formulation can be expressed as:

$$^0k_s = {}^1k_d\,[P_{fr}]$$

or

$$[P_{fr}] = \frac{^0k_s}{^1k_d}$$

This is a surprising result since it implies that the steady state amount of the effector element P_{fr}, $[P_{fr}]$, is independent of wavelength or irradiance of light. (Remember that neither ^0k_s nor ^1k_d depends on light). The only requirement is that the light condition is such that a steady state becomes established. The amount of P_{total}, on the other hand, is dependent on the wavelength of light as shown by the following equation:

$$\frac{[P_{fr}]}{[P_{total}]} = \varphi_\lambda \quad \text{or} \quad [P_{total}] = \text{constant} \cdot \frac{1}{\varphi_\lambda}$$

In words: the ratio φ between the amount of P_{fr} and P_{total} under steady state conditions is a function of wavelength since $^1k_1$ and $^1k_2$ are wavelength dependent. Thus $[P_{total}]$ is a function of φ_λ. The fact that $[P_{fr}]$ is not wavelength dependent at least up to about 730 nm has been used in our research insofar as we establish in our work the photo steady state of the phytochrome system with standard far-red light (a wave band equivalent to 718 nm with regard to the phytochrome system). Since under this kind of light the formation of chlorophyll is negligible, we can study phytochrome-mediated photomorphogenesis without interference of photosynthesis.

The mathematical treatment of the phytochrome model can, of course, be performed on a more sophisticated level. I want to consider only one further aspect. Under steady state conditions the systems analytical treatment of the phytochrome model leads to a linear relationship between the term

$$\frac{P_{total}^{N_\lambda} - P_{total}^{\infty}}{P_{total}^{\infty}}$$

and the reciprocal of the irradiance applied, $\frac{1}{N_\lambda}$ (Fig. 18).

Since P_{total} can be measured in situ with considerable accuracy, the theoretical prediction can be tested by experience. Figure 18 shows that the photometric measurement of the P_{total} term in the cotyledons of the mustard seedling agrees with the prediction and thus confirms the *quantitative* validity of the phytochrome model. The experimental data, obtained with three different approaches, are fitted quite well by the theoretical straight line.

Two further aspects of general significance will now be discussed briefly.

Figure 18. Irradiance dependency of amount of total phytochrome [P_{tot}] present under steady state conditions in continuous standard far-red light in cotyledons of mustard seedling. Term $P_{tot}^{N_\lambda} - P_{tot}^{\infty} / P_{tot}^{\infty}$ is plotted against reciprocal of irradiance N_λ. *Straight line:* theoretical (systems analytical) result; *points:* empirical results obtained with 3 independent experimental approaches. $P_{tot}^{N_\lambda}$, total phytochrome [P_r plus P_{fr}] under steady state conditions at irradiance N_λ, P_{tot}^{∞}, total phytochrome under steady state conditions at infinitely high irradiance [after (6)]

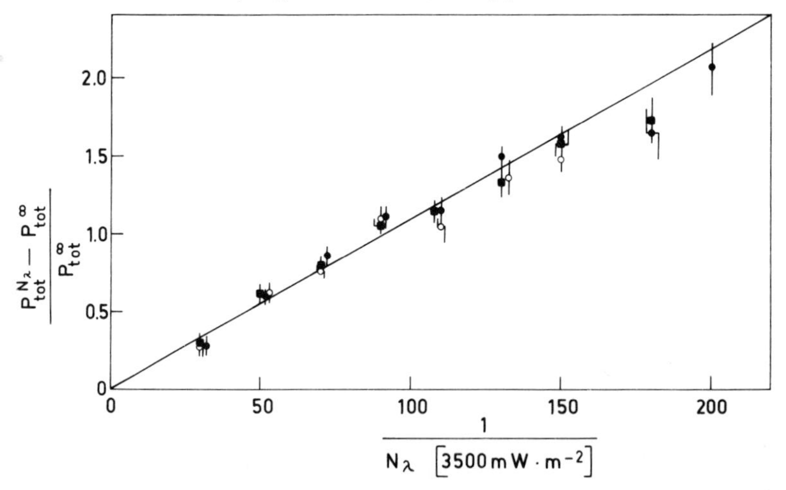

1. We believe that our model derived from measurements with the mustard seedling plainly represents the real phytochrome system as it occurs in the developing higher plant. The model could probably be adapted to the conditions prevailing in other plant species by appropriate changes in the reaction orders and magnitude of the reaction constants (7). In our eyes the model (Fig. 17) is a general proposition, a law-like structure, if the reaction orders and the magnitude of the reaction constants are considered as boundary conditions that must be defined for every particular plant material and light condition.

2. The model (Fig. 17) can be extended. The so-called high irradiance responses of photomorphogenesis, e.g., can only be explained by the model if it is considerably enlarged, i.e., extended by some elements (8). The simplest model of the phytochrome system as pictured in Figure 17 is only adequate for an explanation of photomorphogenesis as long as the law of reciprocity holds ("induction conditions"). We restrict our considerations to this range.

The phytochrome system (Fig. 17) does not show any *cooperative* properties, not even in those cases in which a highly ordered dichroic arrangement of the phytochrome molecules must be assumed (9, 10). The first order rate constants show that in the photoconversions as well as in the destruction process the molecules are used up in the reaction processes in a *random* fashion. The question has been whether or not this is also true for the so-called primary reaction of phytochrome, i.e., for the reaction of P_{fr} with its primary reactant under "induction conditions."

According to our present knowledge this reaction can be represented by the following model:

$$P_{fr} + X \rightarrow P_{fr}\,X \rightarrow P_{fr}\,X' \dashrightarrow \Delta A$$

where X is the primary reactant; X′ is a somewhat modified state of the primary reactant, the combination $P_{fr}\,X'$ being the starting point for the so-called initial action of P_{fr}, which eventually leads to a directly observational change ΔA, called "photoresponse," e.g., synthesis of anthocyanin.

The most important question in this context has been whether or not the huge diversity of photoresponses is based on one particular primary reaction. It has been a wide-spread prejudice among plant physiologists that there is only a single primary reaction of phytochrome. While the reason for this psychologically interesting prejudice is not clear (the theory of photomorphogenesis becomes *more difficult* if the diversity of photoresponses must be explained on the basis of a single primary reaction!), the prejudice has inhibited progress in this matter for several years since it was cultivated by some *influential* researchers in the field. At present it is clear that the notion of a single primary reaction of phytochrome cannot be maintained. Rather, there is *experimental* evidence that there are several primary reactions of P_{fr} that differ essentially, namely in the cooperativity of the $P_{fr}\,X \rightarrow P_{fr}\,X'$ transition.

1. High cooperativity of the $P_{fr}\,X \rightarrow P_{fr}\,X'$ conversion: our work on the control of lipoxygenase (LOG) synthesis in the mustard seedling cotyledons has shown that this control is exerted through an all-or-none reaction. The principle of the response is that synthesis of the enzyme LOG is fully and immediately inhibited as soon as a certain threshold value of P_{fr} is exceeded. Synthesis of LOG is resumed immediately and with full speed as soon as the amount of P_{fr} falls below the threshold value (see Fig. 13). The explanatory hypothesis is that P_{fr} is a ligand of a reaction matrix. The primary reactant X is an integral constituent of this reaction matrix. The reaction matrix possesses the property of performing fully reversible conformational changes with a high degree of cooperativity. The reversible conformational changes are related to the primary reaction of P_{fr} as indicated in Figure 19.

Neither the hypothesis nor the experimental evidence supporting it need to be developed here in detail (11) since the decisive point for our present consideration (significance of the system concept for physiology) can already be made: the high degree of cooperativity of the $P_{fr}\,X \rightleftarrows P_{fr}\,X'$ transition is a *system* property. It is not a property of P_{fr}, but only a property of the *interaction* between P_{fr} and the specific reaction matrix. Figure 12 shows that the effect of P_{fr} on LOG synthesis can indeed be related to the actual amount (level) of P_{fr}; however, P_{fr} can only work with regard to LOG synthesis as long as the organized complexity, including the high cooperativity in the conversion of the reaction matrix, is preserved.

2. No cooperativity of the $P_{fr}\,X \rightarrow P_{fr}\,X'$ reaction: In the case of phytochrome-mediated anthocyanin synthesis in the mustard seedling cotyledons the (P_{fr})-response-curve (dose-response-curve) extrapolates through point zero and is linear over a considerable range (see Fig. 8). This implies that there is no cooperativity in the $P_{fr}\,X \rightarrow P_{fr}\,X'$ transition. The conclusion that the primary reaction in cases (1) and (2) must be different

Figure 19. Simplest formulation of actual threshold reaction in lipoxygenase response. Element X′ is taken from open phytochrome-receptor model advanced by SCHÄFER [8]. It was suggested that receptor X to which phytochrome rapidly binds exists in two forms (X and X′) and that transition $X \rightarrow X'$ is mediated by P_{fr} above threshold [after (11)]

$$P_{fr}X \rightleftarrows P_{fr}X'$$

stable at $[P_{fr}] < 1.25$ per cent	stable at $[P_{fr}] > 1.25$ per cent
no P_{fr} destruction	P_{fr} destruction (^1k_d)
LOG „synthesis" unimpaired	LOG „synthesis" suppressed

can hardly be avoided. It is possible that X and X′ is the same in both reactions, but then the structure of the reaction matrix of which X (X′) is an integral part must be considerably different in both cases. In conclusion, the model for the primary reaction of phytochrome,

$$P_{fr} + X \rightarrow P_{fr}\,X \rightarrow P_{fr}\,X'$$

can be maintained as a general proposition. However, the boundary conditions may vary from response to response, probably due to the attachment of X (X′) to very different reaction matrices.

In the following I summarize some guiding principles for research in physiology that have been illustrated by the examples just described.

1. Research in physiology must be guided by system models that are formulated in the style of systems theory and can thus be quantified.
2. The models must be kept adaptable. This implies that a change of a few parameters in the models permits a description of the situation that prevails in *different* species. This kind of super models would open the way to a general (rather than species or class specific) physiology. The super models must not be trivial. It would not make sense to reach generality by simplification, i.e., at the cost of precision and loss of correspondence to reality.
3. Biological effector molecules (e.g., phytochrome, hormones, inductor substances) can only be recognized in their real significance if they are treated as elements of systems.
4. One and the same effector molecule can act very differently even in the same organism depending on the "organized complexity" in which it enters as an element.
5. In the performance of in vitro studies at least part of the organized complexity will get lost. This implies a loss of (essential) system properties. The decision over the *biological* significance of an in vitro study is, therefore, made in the *physiological* experiment. The art of research to perform significant experiments and measurements with the intact organism, i.e., at full preservation of the organized complexity, will remain the basis of experimental biology.
6. This latter postulate is not in opposition to the tendency to molecularize the *elements* of the system. The system as black box, defined by input and output, is not satisfactory, even after a stochastic refinement. The postulate means, however, that a molecularization of the elements of a system is only significant for experimental biology if the relationship between the elements under preservation of the system properties is studied in a parallel effort.
7. While whole biological theories have rarely been brought in an axiomatic form, smaller areas in biology can be treated successfully in an axiomatic-deductive manner. In physiology, the basis for deduction is system models such as the one in Figure 17, supplemented by boundary

and initial conditions. The deductive treatment introduces an element of clarity and security that is often lacking in those parts of biology that cannot (yet) be formalized, but proceed in a rather loose and informal manner. It is essential, of course, that the models on which the deduction is based represent the real system "out there" with satisfactory approximation. This can only be tested by experience (see Fig. 18).

8. Some philosophers [e.g., SMART (12)] feel that the axiomatic-deductive ideal is nowhere reached in biology. They insist that it is essential for a "theory" to have a logically unified structure. In this strict sense probably no biological theory is really satisfactory. However, the phytochrome example shows that at least *parts* of physiological theories can be treated in accordance with the axiomatic-deductive ideal. Other examples include WOODGER'S approach to axiomatize classical Mendelian genetics (13), including parts of population genetics, or MARY WILLIAMS' effort to axiomatize parts of evolutionary theory (14). Despite these examples there can be no doubt that there is no theoretical biology in the way that there is theoretical physics. It is only within some isolated areas of biology that biological propositions can be arranged in a truly systematic order.

The primary goal of *comparative* biology is the functional and teleological explanation of living systems and their parts. A secondary goal is to contribute to an appropriate classification of living systems and to the understanding of phylogeny.

I have already discussed and hopefully vindicated the *structure* of functional or teleological explanations following essentially the treatments given by RUSE and HULL (15, 16) (see p. 73). I shall briefly recall the major aspects: functional and teleological accounts are explanatory insofar as they state what role a "part" plays within the "whole." "Part" can be any structure, feature, molecule, process, or piece of behavior. The "whole" can be an organism or a group of organisms. ("The function of chlorophyll is to absorb light quanta for photosynthesis"; "The function of flower anthocyanin is to attract pollinating animals"; "The purpose of a proplastid is to give rise to a chloroplast.") In a functional or teleological explanation we assume as a matter of course—and this is an indispensable paradigm—that living systems are "teleonomic" systems operating on the basis of a rigid and optimized program. The necessary information is predominantly coded into the DNA of the organism and released sequentially during ontogenetic development. The optimization of the program has occurred during the evolutionary process by natural selection. The "goal" has been adaptation to the particular environment. It is highly likely that any structure, process, or behavior of an organism will fulfill a positive function with a high degree of efficiency. The seemingly goal-directedness of ontogenetic development (developmental homeostasis) as well as the purposefulness of steady state regulatory mechanisms (homeostasis) can both be accounted for by natural selection. In the case of vestigial organs it is equally likely that they have fulfilled a *positive* function in the *past* history of the species.

The *purposefulness* of an organ or a function is always implied when we ask what is the contribution of a particular structure, process, or behavior to the proper maintenance of the living system? It is not only the purpose for the individual which must be considered, but also the purpose for the species or at least for the population. An organ or a behavioral trait that is purposeful for the population must not be purposeful for the individual.

It is obvious that functional and teleological statements and explanations can only be made against the background of the theory of evolution including adaptation by natural selection, or against the background of a willingly created world, created and optimized by an omniscent and omnipotent creator who has acted consciously according to the scheme of teleologic thought and action. In COPERNICUS' (17) words: ". . . the world machine, created for our sake by the best and most systematic Artisan of all." As scientists of the twentieth century, we adhere to the theory of evolution, which is the most important and most stable paradigm of present biology.

Irrespective, however, of the particular background, functional and teleological explanations are supposed to make the system and its parts comprehensible. If we trust in the paradigms of evolution we must assume as a matter of course that there is no structure without function. As I pointed out previously, the biologist resorts to functional or teleological explanations, not because a *causal* account of a life process could not always be achieved in principle, but because functional and teleological explanations are demanded by the very nature of biological systems and by the type of understanding we seek. Again, our central propensity structure comes heavily into play.

Can comparative biology formulate laws? The answer is yes. Some of the most beautiful laws we know in the sciences are laws of comparative biology. Before I can discuss their nature I must briefly introduce "taxonomy," since comparative biology and taxonomy mutually depend on each other. Comparative biology is mainly concerned with the comparison of the properties of taxa. On the other hand, the creation of the construct "taxon" and the association of a particular organism to a particular taxon within a particular category depends on description *and* comparison. The laws of comparative biology further depend on the validity of the principle of homology, which must be considered to be one of the most powerful intellectual inventions in science (18).

From a logical point of view the laws of comparative biology must be considered as coexistence laws (see p. 63) with a very high predictive power. For example, the general proposition: "In the Spermatophytes the contents of the mature embryo sac is homologous to a female gametophyte" is a coexistence law, valid for all spermatophytes. As soon as one identifies something as a spermatophyte, one can make a large number of forecasts about its morphology, anatomy, cell physiology, reproductive processes, chemical constituents, etc., including the forecast that the female gametophyte of the organism will be found in the mature embryo sac. As already indicated, this statement is not independent of the validity of other principles; it implies at least the validity of the homology concept in comparative biology. However, it does not imply the validity of the concept of phylogeny or of any theory regarding the evolutionary *mechanism*. Therefore, the laws of comparative biology are insensitive to changes in the theory of evolution as long as they are formulated as coexistence laws implying only the validity of the concept of homology.

Another example to make a further point: "In the Tracheophytes mature micromeiospores produce male gametophytes." Again the predictive value of this coexistence law is extremely high. If we identify among the Tracheophytes a particle as a mature microspore, we know in advance that the germination of the particle will inevitably lead to a male gametophyte. The laws of comparative biology, implying the homology principle, are restricted laws with a limited scope. They are only valid for certain taxa of organisms, such as Spermatophytes, Tracheophytes, or Vertebrates, which can be treated in accordance with the homology principle. Within these limits these coexistence laws are very reliable. Therefore, we may depend on them not only in physiology, but also in biological technology.

I have indicated already that it has been a goal of comparative biology to contribute to our understanding of phylogeny and to an appropriate classification of organisms. From the point of view of comparative biology this might be regarded as a secondary goal. However, this kind of subsumption does not mean that classification (taxonomy) per se would be a secondary science. Rather, the tendency for classification is a very primary, innate drive in man, part of our central propensity structure. Since science must be considered to be a cultivated, refined, and intellectually controlled venture in accordance with our innate propensity structure, it is understandable that classification of observable objects has been fundamental in most scientific disciplines. Classification is thus one aspect of the very human activity to organize logically the real world using a set of "constructs." In this undertaking we follow our deep faith (our "foreknowledge") that nature is an orderly system. I have mentioned on previous occasions that our foreknowledge about the real world includes order and consistency, and we have already discussed the fact that we are able (without too much reflection) to think in observable "constructs," in entities. A tree is a construct, a flower is a construct, a species is a construct. A tree, a flower, a biological species are real in the sense that an observational construct is real (see p. 37). In any case the construct "biological species" is not a scientific invention. Totally illiterate aborigines recognize a biological species as readily as the experienced naturalist does. Even MAYR (19) admits that: " . . . primitive natives in the mountains of New Guinea will distinguish the same kinds of organisms as, quite independently, does the specialist in the big national museums." Thus the *principal* recognition and the usage of the construct "species" seem to be firmly anchored in our foreknowledge about the world. However, the explicit, precise *definition* of what a species is has created great difficulties. Every experienced taxonomist is aware of this. My teacher in taxonomy, the paleobotanist Walter Zimmermann, once answered our question about the nominal definition of the term "species" as follows: "A biological species is what an experienced taxonomist considers to be a species." This statement (which has also been attributed to the British ichthyologist Regan) certainly exaggerates the unavoidable subjectivity of taxonomy, but even the most formalized methods of phenetic taxonomy cannot eliminate this subjective element completely, which includes experience, intuition, taste, and even genius. I do not think that description of species and higher classification of species can in fact be freed from the individual bias of the experienced taxonomist, although all effort should be undertaken to minimize the inevitable risk of a subjective judgment, in particular in the genealogical approach.

According to MAYR (20), biological species are "groups of actually or potentially interbreeding natural populations which are reproductively isolated from other such groups." While this nominal definition is attractive, it obviously does not readily apply to those plant "species" that do not practice sexuality or where interspecific fertile hybrids are not only produced under experimental conditions, but also occur in the natural habitat.

Before we discuss the assumptions and inherent difficulties of taxonomy further, I want to remind you briefly what biological classification means. The construct species and the formal rules of classification are entirely different things. As I pointed out previously, the construct species must be considered to be an inborn construct; the straightforward formal rules of classification, on the other hand, were provided by Linnaeus. It is not the construct of "species" and it is not the set of formal rules of classification that creates the problem. Rather, it is the next step, the interpretation of the Linnaean structure from the point of view of genetics and evolutionary theory, which is laden with assumptions and difficulties.

Linnaeus suggested that organisms be classified hierarchically. You are all familiar with his basic concept, to define an ordered set of classes referred to as "categories." Every category has as members classes referred to as "taxa." Organisms are members of taxa. The important point in Linnaeus' concept of classification is that a particular organism can become a member of only a single taxon in each category. In the categorical hierarchy (as we presently use it) each organism must belong to a minimum of seven taxa in seven different categories of increasing "rank." With regard to the white-seeded mustard plant (our favorite subject in research on developmental genetics) the categories and the taxa are the following:

Category	Taxon
kingdom	Plantae
division	Magnoliophyta
class	Magnoliatae
order	Capparales
family	Brassicaceae
genus	*Sinapis*
species	*Sinapis alba*

I want to emphasize that this formal structure does not depend on any genetic or evolutionary knowledge or concept. This is a great advantage of the Linnaean system from the logical point of view; on the other hand, it is clearly a disadvantage with regard to the *empirical content* of the Linnaean system: to what extent—this is the crucial question—does the Linnaean system of classification represent genealogical relationships? However, this question must be kept clearly separate from any consideration about the usefulness of the Linnaean system for the purpose of *formal classification*. It is evolutionary taxonomy that tries to correct the "artificial," i.e., primarily logical, Linnaean system by the import of genetic and evolutionary knowledge, and this effort creates difficulties (15).

While the biological species is believed by most biologists to be real (in the sense an observational construct is real), the higher categories of Linnaeus' classification are clearly *theoretical* constructs whose "reality" is much more difficult to establish. Usually, these categories are considered to be arbitrary.

For the geneticists among the evolutionary taxonomists, the dominant consideration is the similarities and differences of the gene pools of populations (species) they have to classify. As MAYR (21) puts it: "When an evolutionary taxonomist speaks of the relationship of various taxa, he is quite right in thinking in terms of genetic similarity, rather than in terms of genealogy." The genealogists among the evolutionary taxonomists emphasize the organism's phylogenetic history. As SIMPSON (22) says: "It is preferable to consider evolutionary classification not as expressing phylogeny not even as based on it, . . . but as consistent with it. A consistent evolutionary classification is one whose implications, drawn according to stated criteria of such classification, do not contradict the classifier's view as to the phylogeny of the group." The two points of view are not mutually exclusive, of course, since it is generally agreed among evolutionary taxonomists that phylogenetic diversity is based on genetic diversity. The somewhat different views expressed by Mayr and Simpson obviously reflect the different interests: SIMPSON who emphasizes phylogeny in classification is a paleontologist; MAYR, who emphasizes genetics, is a neontologist (15).

To what extent is the evolutionary taxonomist's approach, as pointed out by Mayr, consistent with *developmental genetics?* Developmental genetics is that part of modern biology that investigates the causalities involved in the expression of an organism's genes in the course of ontogeny. We recall at this occasion that living systems are continuously developing systems and that they can only be understood if we respect this feature. Any static description is a priori inadequate. The living system can only be represented adequately by the *time course of ontogeny*, not by a cross-section somewhere in ontogeny.

The evolutionary taxonomist (as every taxonomist) works on the observational level. He will use every available character, morphological, physiological, chemical, and even ethological. Statements about theoretical entities such as genes are always inferred statements. The taxonomist must infer statements about genotypes from statements about phenotypes. He must infer genetic differences from observational differences between organisms. Even though we became accustomed to talk about gene pools and frequencies, it is the phenotype that we observe. The problem is that the relationship between the observational level (traits) and the theoretical level (genes) is neither simple nor unambiguous. This aspect will now be considered from the point of view of the developmental geneticist.

1. In the lecture on scientific language and terminology I criticized the fact that the bridge principles implied between the observational level (traits) and the theoretical level (genes) are mostly considered to be a matter of course. Evolutionary taxonomists talk quite lightly about genes, ignoring not only the direction of inference but also the inherent difficulties with regard to the bridge principles. As an example, MAYR (19) writes: "The reproductive isolation of a biological species, the protection of its collective gene-pool against pollution by genes from other species, results in a discontinuity not only of the genotype of the species, but also of its morphology and other aspects of the phenotype produced by this genotype. This is the fact on which taxonomic practice is based."

2. The unit of taxonomy is a population of organisms. The members of any population show variation with regard to every trait we measure. We assume that the total variation is composed of two parts, genetic variation and environmentally determined variation, but we do not know—without sophisticated experimentation—the contribution of either type of variation to the total variation of a trait in a given population. From developmental genetics we know that there are traits (characters) whose expression is hardly influenced *specifically* by the environment (developmental homeostasis). The variation of such a trait in a population is

predominantly genetic variation (3). On the other hand, there are characters of the same organism whose quantitative and sometimes even qualitative expression is *specifically* dependent on particular environmental factors (3). The phenotype–genotype relationship is thus specific for every trait and must be determined experimentally for every trait if a quantitative statement is desired. A rigid genotype–phenotype relationship (a given genotype gives rise to the same phenotype under a wide variety of environmental conditions) has been assumed as a matter of course in most of classical Mendelian transmission genetics, population genetics, and evolutionary taxonomy. Developmental genetics has shown, however, that the relationship between genes and traits will depend in some traits of an organism on *specific* environmental factors such as light, photoperiod, gravity, or temperature while other characters of the same organism show indeed strict developmental homeostasis (3). Control by the environment and rigid developmental homeostasis can occur sequentially during the development of one and the same trait. As an example, flower *induction* in an obligatory (qualitative) short-day plant depends entirely on the appropriate photoperiod. However, once induction has occurred, the actual *development* of the flowers is characterized by rigid developmental homeostasis (3).

With regard to vegetative characters of a plant, it is obvious that the *dimensions* of organs such as leaves or internodes will depend heavily on the particular environment, for instance on the light factor, whereas the nature of patterns, e.g., in phyllotaxis, cannot be influenced specifically by any factor from the normal environment of a plant (Fig. 20).

From the point of view of the developmental geneticist it is mandatory that the taxonomist consider predominantly those traits that are characterized by rigid developmental homeostasis in order to minimize the variability introduced by environmental fluctuations. This explains the traditional preference in plant taxonomy for those traits associated with flowers. The experienced taxonomist "knows" intuitively or by inductive generalization that he must give certain characters greater "weight" than other characters. MAYR (19) writes: "The scientific basis of *a posteriori* weighting is not entirely clear, but difference in weight somehow results from the complexity of the relationship between genotype and phenotype. Characters which appear to be the product of a major and deeply integrated portion of the genotype have a high information content concerning other characters (which are also products of this genotype) and are thus taxonomically important. Other kinds of characters, . . . as well as superficial similarities, convergences, and narrow adaptations, have low information contents concerning the remainder of the genotype and are thus of low value in the construction of a classification." It is developmental genetics that can rationalize and justify these intuitive assumptions of the evolutionary taxonomist and protect them against objections from the advocates of phenetic taxonomy.

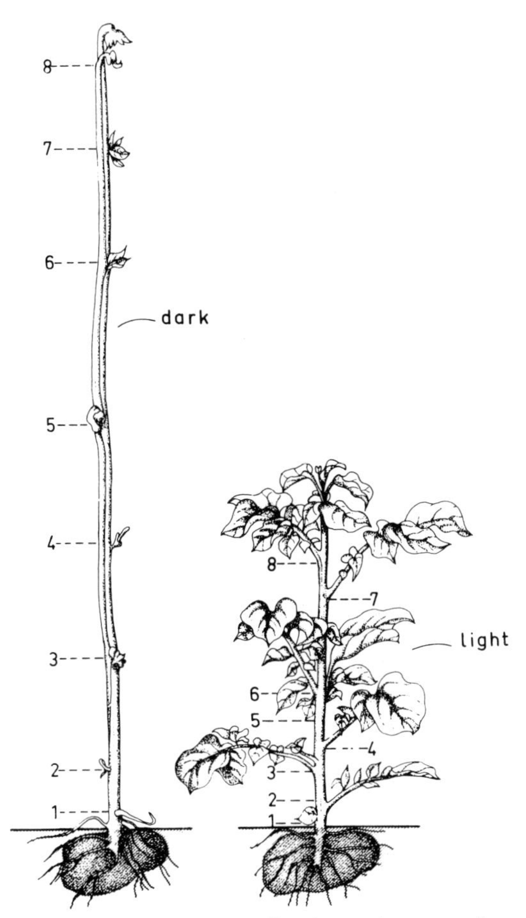

Figure 20. Two genetically identical potato plants *(Solanum tuberosum)*. Nevertheless, dark-grown *(left)* and light-grown *(right)* plants differ greatly. However, pattern of phyllotaxis is obviously not influenced by the presence or absence of light [after (3)]

3. If two organisms have the same genotype and develop in the same environment, they will *necessarily* develop the same phenotypic characters. This general statement can be inferred from the principle of causality and can thus be regarded as an example of the deduction of a particular law from a more general principle. In practice, the statement is the basis for the circumstance that we infer from a phenotypic mutation the occurrence of a mutation at the gene level.

The foregoing statement that two organisms that have the same genotype and develop in the same environment will develop the same phenotype is not symmetric in the sense that identity of *phenotype* necessarily implies identity of genotype. While this state of affairs was recognized even by classical Mendelian genetics, the problem inherent in the phenotype–genotype relationship has become much more severe in recent years. The existence of selectively neutral mutations that can be measured at the protein level but that do not become expressed at the phenotypic level[6] can hardly be doubted any more. If neutral mutations are possible, then organisms with considerably different genotypes may show precisely the same phenotype. Therefore, if two organisms show precisely the same phenotype, it may not be concluded that they have identical genotypes. Considering evolution, the possibility of neutral mutations leads to the conclusion that the principle of natural selection does not apply at the level of genes (DNA) or proteins in the way it does at the phenotypic level (23).

4. Sometimes a very slight genetic difference can lead to gross morphological differences. An example is the change of the dorsiventral flower of *Linaria vulgaris* or *Antirrhinum majus* to a radically symmetrical flower by a simple mutational step (Fig. 21). The two populations of *Linaria vulgaris* we have documented (24) are genetically very similar but morphologically very different, at least in the structure of the flower. We must conclude that the notion of genetic similarity or dissimilarity cannot be used as an ultimate court of appeal in taxonomy, however valuable this notion might be in comparing established, "good" species within a limited branch of the evolutionary tree with regard to similarity or dissimilarity of DNA, e.g., in man and the apes or in rat and mouse.

[6]The phenotype consists of those traits that are relevant for selection.

Figure 21. Inflorescences of *Linaria vulgaris*. *Left:* normal specimen with dorsiventral flowers; *right:* atavistic mutant with radially symmetrical flowers [after (24)]

In conclusion, species must be conceived and defined on the *observational* level. A common gene pool can be inferred from the existence of a species, but the common gene pool can hardly be used to establish a species. A similar conclusion must be drawn with regard to Simpson's evolutionary species. It may be inferred that an actual biological species is a temporal cross-section of an evolutionary species, but this is a kind of inference that already implies the existence of the species on the observational level.

As everywhere in science, the indiscriminate use of observational and theoretical terms without explicitly mentioning the bridge principles, will create confusion. It was hoped by some biologists that the highly operationalized and formalized phenetic taxonomy could overcome these difficulties of evolutionary taxonomy (25). However, as things have developed it seems doubtful that phentic taxonomy really offers a superior alternative.

The phenetic taxonomists consider quantitatively overall physical resemblance (similarities and differences), called "phenetic" resemblances, between the members of a group of organisms (sample) with the highest possible degree of objectivity while ignoring genetic or evolutionary arguments. The goal is to erect a taxonomic system that is "objective" and allows trustworthy predictions.

Since I agree with DELEVORYAS (26) that "if a system is not phylogenetic, it is really meaningless"; and since a critical evaluation of phenetic taxonomy has appeared recently in RUSE'S (15) excellent treatise, I will not enter this field of debate. My feeling is, however, that the numerical methods of phenetic taxonomy should be made useful in evolutionary taxonomy in providing better means for a quantitative description of the distribution function of the most relevant characters of an intuitively conceived species. The phenetic taxonomist's effort to achieve objectivity and predictability by conforming to the principle of operationalism is not a bad thing, even though the theoreticians among the phenetic taxonomists seem to overestimate the degree of objectivity they in fact obtain.

However, the principal difference between evolutionary and phenetic taxonomy lies in the differing interests, or in the different goals. If we want the actual classification of organisms to reflect genealogical relationships, we can accept the approach of phenetic taxonomy as a means of description of resemblance, but not as the goal of taxonomy. "Subjectivity" is the major accusation advanced by phenetic taxonomy against evolutionary taxonomy. This charge of wrong-doing must be taken very seriously, in particular since the leading theoreticians of evolutionary taxonomy are aware of the partly subjective nature of their classificatory statements. SIMPSON (27) writes: "Taxonomy is a science, but its application to classification involves a great deal of human contrivance and ingenuity, in short, of art. In this art, there is leeway for personal taste, even foibles, but there are also canons that help to make some classification better, more meaningful, more useful than others."

With regard to "objectivity" I have repeatedly emphasized the point that all scientific undertaking depends on inborn foreknowledge as well as on our inherited central propensity structure (see Lecture on Epistemology and Evolution). All science is selective in the sense that it weighs the facts of experience and that it chooses *certain* hypotheses, thus bypassing the multitude of *possible* hypotheses. Thematic presuppositions have always been part of scientific discovery (28). This kind of statement does not imply permission for uncontrolled phantasy or deliberate violation of the consensus principle. The statement means, however, that even objectivity in the strict sense of the consensus principle does not guarantee that we are unbiased or neutral against the real world "out there."

In evolutionary taxonomy (at least with regard to higher taxa) the consensus principle is strictly obeyed *with respect to the rules* that ought to be applied. The individual taxonomist's degree of freedom is strictly limited in this regard. And that is probably all that can be demanded.

10th Lecture

Tradition and Progress in Science: The Notion of Paradigms

Science is a social activity. The scientific community, past and present, represents the social dimension (1) of the scientific enterprise. The scientific community is commonly regarded as the internationale of fellow professionals who scrutinize each other's work, correct it, and carry it on. This latter point implies "tradition," in practical terms, the instrumental, intellectual, and moral training of the student as part of his preparation for future membership in the scientific community. In the first lecture (Prologue) I discussed the reasons why the scientific internationale is a purely professional one and carries no political weight.

There are two types of scientific communities depending on the forces that keep the communities together. The global community of all natural scientists is operationally defined by the allegiance of its members toward the normative code of science, the intrinsic values of science. Since this point is essential for the function of the scientific enterprise, the ethics of science deserve a thorough consideration in a separate lecture (see next chapter).

A great many scientific communities of limited scope and membership consist of the practitioners of a particular scientific discipline. We all know the main scientific professional groups such as physicists, chemists, botanists, astronomers, and the more concrete groupings, professional societies. However, communities in this sense exist at numerous levels, including the official academies and the unofficial "invisible colleges" around some outstanding person with a high degree of scientific leadership and charisma. These scientific communities play a role (sometimes disagreeable) as restrictive guilds or select fraternities since membership depends on election or some other kind of adoption. More recently, scientific communities have become established that are defined by instrumentation (Electron Microscopical Society), by particular techniques (Spectroscopical Society), by environmental factors (Photobiological Society), or by practical goals (Plant Breeding Society). It is the rule that an individual scientist, at least if he is above average in capability, will belong to several such groupings, either simultaneously or in succession. KUHN (2) has emphasized the point that scientific communities, consisting of the practitioners of a scientific discipline, are in fact the units that produce and validate scientific knowledge. He has further suggested that the groupings are kept together by a disciplinary matrix, in particular by *paradigms* that are shared by the members of such groups.

According to Kuhn, "normal" science (in opposition to "revolutionary" science) usually operates within the framework of existing theories, techniques, and behavioral patterns. The loyalty toward tradition is strongly developed; the tendency to overthrow tradition is weak. Normal science is defined by KUHN (2) as "research firmly based upon one or more past scientific achievements, achievements that some particular scientific community acknowledges for a time as supplying the foundation for its further practice." Such achievements are called "paradigms." Our obedience or allegiance toward paradigms expresses itself in the selection of problems, in the selection of methods, and in the choice of intellectual concepts. It would be an illusion to believe that any one of us can escape the directing influence of the paradigms we grew up with. However, the important point is that we retain enough intellectual power and personal courage to overthrow paradigms in favor of new conjectures if there is a really good reason to do so.

Paradigms are the essence of scientific tradition and instruction. It is the study of paradigms that prepares the student for membership in the scientific community. A particular scientific community is thus a grouping of people who share the same paradigms as directives for their behavior as scientists. Paradigms are more than exemplars or models. They include a certain line and even a culture of thought. Paradigms also include a given set of "constructs" or, generally speaking, a certain way to look at things. As members of a scientific community we must see the same things when confronted with the same situation. We have already mentioned the "cell" as a construct fully established in biological tradition. Kuhn would probably consider constructs such as the gene or the quantum of action, which are deeply rooted in the tradition of a scientific discipline, as paradigms. According to Kuhn the acquisition of paradigms can be regarded as a sign of maturity of a scientific discipline.

The most powerful and at the same time indispensable paradigm in biology is the theory of evolution. The presupposition of pre-Darwinian biologists that nature represents an ordained system (ordered by God) has been replaced by the concept that every living system is the product of evolution. This has changed the view of living nature (or, the way we look at living nature) completely, although not necessarily the scientific results. As SIMON (3) puts it: " . . . classification carried out on this basis has preserved many of the Linnaean groupings, in spite of the radical difference in conceptions of how these groups came into being. What seemed a rational and convenient system of naming and classifying also served as a way of representing the sequence of events that constitute evolutionary history." However, "part of the achievement of evolutionary theory consists in the fact that it offers an explanation both of the occurrence of groups of organisms sharing a number of common attributes *and* of the *failure* of nature to exhibit totally discrete types. Systematics based on descent and variation account both for the discontinuities and for the continuities that are observed within the range of terrestrial organisms" (3).

As we discussed previously (see p. 116), the theory of evolution (in a Darwinian sense) has become firmly the basis for functional or teleological explanations in biology. It is implied in every biological explanation that living systems are highly adapted in morphology, development, function, *and behavior* and that this adaptation is the result of a Darwinian evolutionary process. Any deviation from this paradigm in the context of a scientific discussion would be regarded as ridiculous or at least as irresponsible toward the unexperienced student in the field of biology. This is true also for the combination of molecular biology and evolutionary theory. A newly emerging field of "molecular evolution" (analyses of the evolution of enzymes and other proteins) is based on the techniques and data of molecular biology and (as a matter of course) on the paradigms of evolutionary theory.

The practicing scientist accepts the existence of a multiplicity of paradigms as an obvious and "natural" thing. In the daily work at the laboratory bench or at the desk one cannot constantly question the existing constructs and theories (or, more generally speaking, the intellectual framework within which one is working). Biologists rarely spend much time arguing out the evidence for the existence of genes or discussing the empirical basis for the first law of thermodynamics. They take these paradigms for granted. Chargaff, I guess, did not reflect about the validity of the concepts of atomic and molecular weights when he performed the inquiries that lead to "Chargaff's rules" about the composition of DNA (the number of adenine molecules equals the number of thymine molecules, whereas the number of guanine molecules equals the number of cytosine molecules).

The acquisition of paradigms by a person is an act of mental relief. It frees the mind for high level research of the kind which KUHN (2) calls puzzle-solving. KUHN compares normal science, proceeding under the guidance of a paradigm or a set of paradigms, with puzzle-solving: it requires high skill; solving a puzzle does not *necessarily* have much value for anybody else; the *existence* of a solution is assured in advance and the kind of solution that is wanted is known. If I give a sample of crystalline anthocyanin to an experienced organic chemist and ask him to determine the structure, I know the *kind* of solution of the problem and I further know that a solution exists.

The notion of puzzle-solving is not to be taken as a *negative* metaphor. Rather, in a mature science a high value is accorded to puzzle-solving abilities.

"Normal science" and "puzzle-solving" do not imply the notion of a straightforward process, governed only by evidence and logic and virtually free from intellectual fights and irrational elements. *All* novel ideas and data, in "normal science" as well as in "revolutionary science", will meet great resistance that can only be overcome gradually. Even a *minor* step can encounter strong and embittered reluctance if it contradicts an alternative concept that somebody else has staked his reputation on. Progress in "normal science" as well as in "revolutionary science" is made by man and thus involves all features of the nature of man, whether we like them all or not.

Another point in "normal" science, which is seriously underestimated by Kuhn, concerns the willingness and ability of the scientist to consider the possibility that paradigms must be revised or exchanged. Scientists in mature fields are certainly conservative. Their reluctance to accept novelties—on the level of constructs and theories—is legitimate. It would be very unwise to treat every set of discordant data as a falsifying argument against a theory. However, if the existence of *data* that are not compatible with existing theory has become an inescapable fact, no experienced scientist would hesitate to question existing theories. The decisive element is *facts* even though these facts have been obtained within the intellectual and methodological framework of existing paradigms. In science, objective data are stronger than paradigms. However, the objectivity of the data must be proved with the highest degree of confidence.

I want to discuss now some crucial problems concerning "normal" science proceeding under the guidance of a set of paradigms.

1. According to Kuhn, authorship of books ceases to be a principal sign of professional achievement in fields governed by paradigms. "The scientist who writes (a book) is more likely to find his professional reputation impaired than enhanced" (2). In a mature field ruled by accepted paradigms the highly specialized and technical research communication becomes the vehicle to spread information through the community of fellow experts. This has usually the consequence that subcommunities originate (e.g., the *Drosophila* group or the phytochrome group in developmental genetics) with relatively few members and with a strong tendency for intellectual inbreeding and self-flattery. As a rule, superior books or integrating monographs are indispensable for reuniting from time to time the diverging subcommunities. This kind of book is usually written by a strong personality.

2. Research in a mature field with accepted paradigms shows a strong tendency to proceed according to the motto: science for science sake. As already pointed out, Kuhn compares this kind of research activity

with puzzle-solving, and JEVONS (4) makes the criticism that "while a paradigm or set of paradigms may be very successful in guiding puzzle-solving, it may also orient scientists away from socially important problems that are not currently reducible to puzzle form. Scientists tend to tackle problems they *can* solve rather than problems that "need" solving from society's point of view: to investigate fundamental particles or build up complex heterocyclic molecules rather than find a cure for cancer." In my opinion this kind of critique—although quite common and to be taken seriously—misses the point completely. Certainly, the academic freedom of science has led to much "high-sounding humbug" [to use JEVON'S (4) expression]; however, the fact that a closed discipline of the natural sciences is based on truly reliable paradigms and has a wealth of equally reliable laws and facts available by far compensates for some too esoteric adventures. It is the set of established and approved paradigms together with the reliable empirical laws and facts of a mature scientific discipline that can serve *any time* as a trustworthy basis for rapid technological innovations. None of us knows which kind of sophisticated technology we will need tomorrow simply to survive. However, if we have at our disposal the scientific information of a whole discipline, intellectually organized in paradigms, laws, and facts, the technological innovator can hardly fail in solving any problem that can be solved by technological means. From the point of view of human survival we must champion pure, basic research for the reason that it is the only means of providing for a *flexible* technology. The network of scientific information must be at hand when the need arises. Mission-oriented research must fail if the soil has not been prepared by curiosity-oriented research (see Fig. 22).

3. Editors of leading scientific journals and the reviewers they engage to examine the submitted manuscripts are in a particularly difficult position. They must demonstrate some allegiance toward the established paradigms and thus impede unjustified attempts to violate or to overthrow the paradigms (in other words, they must counteract any premature "scientific revolution"); on the other hand, however, they must remain agile and vivid enough not to decline those papers that possibly lead a field to new frontiers.

There are many reports (tales and fairy tales) about editors who failed to maintain the right balance between being conservative (favoring "normal" science) and being progressive (permitting a bit of "scientific revolution" once in a while). Since most of these criticisms stem from retrospect, they are rarely fair. Failure in this matter is not always an indicator for mental rigidity. While it is indeed hard to find and to keep on the small path between justified reluctance against surprising novelties and narrow conservatism, a rather clear distinction can be made between being cautious and being biased. The decisive point in the evaluation of a paper is the validity of the empirical facts *and* the internal consistency of the argument. Provided the facts are clear-cut beyond reasonable doubt and the internal

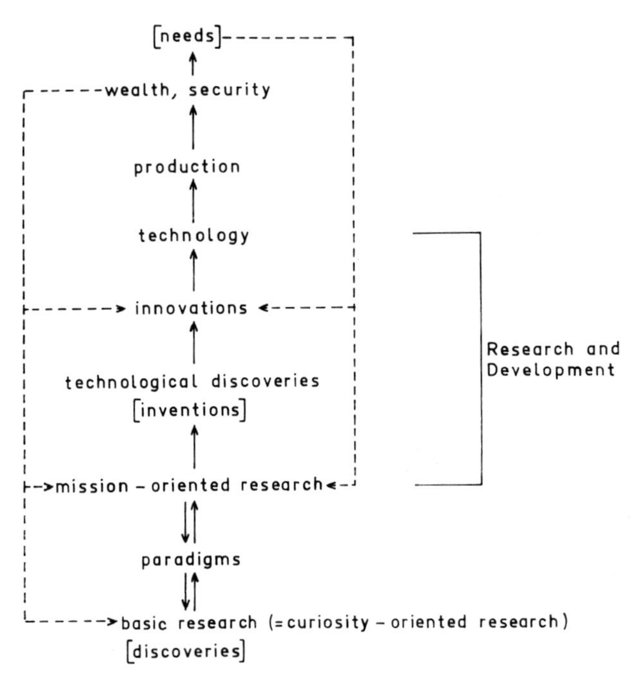

Figure 22. Principal relationship between basic scientific research, technology, production, and wealth. Dashed lines indicate strong feedbacks. Details appear in text

consistency of the argument is given, even the most surprising discoveries should be accepted irrespective of the prejudice or taste of the responsible editor. The knowledgeable editor will pick the appropriate expert referee to judge the merits of a manuscript. That method of choosing referees has played an essential and effective role in the development of science. However, the problem is that the system always depends on individual judgements of quality, and any individual can miss the point.

An editor not only bears the responsibility for weal and woe of a journal (a timid and narrow-minded editor will hardly represent a highly prestigious journal); he is also responsible toward the scientific community to keep out the fancy paper of the pseudoscientific adventurer, but not to miss the short communication of the genius, even though it might be poorly written.

4. Education in science is largely the enterprise of passing on paradigms to the members of the next generation and of teaching them how to use the paradigms to achieve the goals of science. The paradigmatic canon not only includes constructs, laws, and theories, but also intellectual and material techniques and norms of behavior. Education in science is a dialectic enterprise insofar as the desire to achieve a perpetuation of paradigms must be reconciled with the equally important aspect of creating

in the young scientist a critical attitude, a readiness for creative thinking and even for scientific revolutions. The motto of education in science could be; be committed to existing paradigms, but at the same time be committed to discovering new patterns of thought.

Education in science, i.e., teaching and learning of behavioral patterns and paradigms, is similar to teaching and learning in other fields of human life. As LEOPOLD noticed, "the budding scientist may make a strong set of behavioral fixes based on his earliest science experiences in graduate training . . . the same early fixation of behavior could be expected for any profession or craft which involved an extensive apprenticeship" (5). The imprinting influence of a major professor on the budding scientist is easy to discover, from speech mannerism and gestures to the loyal defence of the major professor's scientific hypotheses (5).

The *psychological* component of our educational system is far from perfect. I agree with LEOPOLD'S analysis (5): "Out of the generous number of people trained to the Ph.D. in science, relatively few remain active in research, making for a low ratio of people trained to those active as professional researchers. I feel that this inefficiency is due in large part to the failure of the training system to give young students an enthusiasm for entering the competition and being willing to take the buffeting that is a natural part of the aggressive and competitive interactions with other scientists." We must teach students as early as possible and with much more persuasiveness than usual that competitiveness is a major *positive* factor in bringing about good science and that an *enjoyment* of participating in competitive interactions with professional colleagues is an indispensable part of the scientist's life.

5. The scientific communities cannot afford a serious break in tradition. Even in periods of rapid changes in constructs, styles of thinking, and experimental techniques, a solid core of paradigms, normative behavioral rules, and proved institutional structures must be maintained. The scientific community is very fragile and vulnerable. It constantly requires stabilizing elements. Its existence is not compatible with anarchy or chaos. While it is no bad thing to question and reexamine fundamental beliefs and assumptions, we must do it in accordance with the scientific code of conduct, the ethics of science.

6. According to KUHN (2) a paradigm shift constitutes a "scientific revolution." Indeed, when a paradigm or a set of paradigms is renounced, the world looks different. However, as KUHN (2) notices, "The decision to reject one paradigm is always simultaneously the decision to accept another, and the judgement leading to that decision involves the comparison of both paradigms with nature *and* with each other." Apparently it is an innate drive in human nature to replace obsolete paradigms by new ones. A horror vacui with regard to paradigms is indeed a characteristic of human nature. This is obvious in *all* human life: it has been amazing to watch our rebellious youth in Europe immediately replacing declined

cultural paradigms with new ones, even though the new ones look immature and miserable: le roi est mort, vive le roi! The crucial question of whether or not the new paradigms are really superior to the old ones, can only be answered in retrospect.

The Copernican revolution is a favored example for a dramatic change of paradigms: Ptolemy placed the earth at the center of the universe, Copernicus, the sun. Both looked at virtually the same data. Each system is perfectly logical and self-contained. Copernicus never ventured to give preference to his own system. In retrospect we notice that the Copernican revolution really occurred. In KUHN'S words: "The very ease and rapidity with which astronomers saw new things when looking at old objects with old instruments may make us wish to say that, after Copernicus, astronomers lived in a different world. In any case, their research responded as though that were the case" (2).

The transition from classical to molecular genetics, which most biologists might consider to constitute a paradigm shift, is another interesting case history insofar as the central paradigm in the new set was soon called a "dogma" (6). This "central dogma" of molecular biology includes the double helix model of DNA, the "autocatalytic" function of DNA in serving as a template for the synthesis of replica DNA chains, and the concept that genetic information, piled up in the nucleotide sequence of DNA, will be transcribed into a nucleotide sequence of RNA and translated into amino acid sequences of protein whereby the flux of information occurs only in the nucleic acid → protein direction but never from protein to nucleic acid. The central dogma is, of course, not a dogma in any ideological sense. It has not been intended—as CHARGAFF (7) feared—as a starting point for a new kind of normative biology that commanded nature to behave in accordance with the models. "Central dogma" is just a fancy (but *psychologically* interesting) name for a biological principle that is open to refutation any time. Actually, at least the *mechansim* of DNA replication in vivo is still open for debate and largely obscure.

Some sincere critics, in particular CHARGAFF, have blamed the avant garde of molecular biology for not being in line with the proved tradition of scientific inquiry. "When molecular biology appeared on the scene, . . . , the publicity machines were all in position, and it was time for the saturnalia to begin in full force" (7). I leave it open whether or not this bitter argument is really justified. It could be that the creation of some new set of paradigms at the present state of science and society must *necessarily* be correlated with terminological fancifulness, unusual snobism, and even with the neglect of established rules of behavior. However, CHARGAFF insists that "many of the great constructions of our time—existentialism, structuralism, transformational grammar, the central dogma and some other sloganized tenets of molecular biology—have all looked, from their very beginning, somehow shoddy and overblown" (7).

7. A paradigm shift, and even a minor change in the set of estab-

lished paradigms, is often followed by the super dominance of a particular approach. This may cause very serious and largely irreparable damage to some scientific disciplines. As a sad example, the rise of molecular biology has led to the virtual extinction of whole fields of traditional biology. The "noble study of botany" [to quote CHARGAFF (7)] has been all but banished from many universities. The diversity of methodological and intellectual approaches in experimental biology has been narrowed down to the use of high-speed centrifuges and scintillation counters. A still graver risk for the future of biology as a diversified discipline is the very "modern" anthropocentric attitude. This is a basic paradigm shift in biological sciences leading to adverse consequences on all levels of the scientific enterprise. It is a serious signal for the end of the free inquiry and for a severe constriction of the human mind if research in plant development must be justified toward the granting agency as a constribution to cancer research *in man*. While it is indeed probable that basic research with plants will contribute to the final understanding of the phenomena of cancer, it is not only a loss of diversity, but also a loss of *dignity*, if we can no longer afford to tell frankly that we are interested in studying plants for the sake of genuine knowledge *about plants*. The enforced anthropocentrism in present-day research and teaching is the greatest menace for the future of biology as an autonomous science.

Anthropocentrism in science is by no means to save our endangered world. It is true, of course, that we must understand man in terms of solid science. However, we also must understand and appreciate the rest of the universe in order to make our knowledge about man really fruitful. As I pointed out in the introductory lecture, it has been the autistic anthropocentric attitude that has driven man to the verge of extinction.

8. What does *progress* mean in science? The notion of progress always implies improvement, higher quality. A purely quantitative accumulation and compilation of largely unrelated data should not be considered as "progress." Rather, "a record of steady but pedestrian work over a period can finally convince observers that a field is irremediably boring" (8). Moreover, we all know that there is a category of "pointless publications" or what RAVETZ has called "shoddy science," i.e., immature, ineffective, or rotten fields of inquiry. As RAVETS puts it: "I have explained how under certain circumstances it is possible for published reports of research to contain no substance, but to consist of unsound data, interpreted by an incoherent argument, which leads to a vacuous conclusion. This can happen in spite of scientists' good intentions, in an immature field; and it is brought about by a corruption of standards in the case of shoddy science in an established field. Whatever the cause, we are thereby reminded that publication does not equal progress" (8). However, progress in science is a real thing at least in many fields. *There are* useful internal criteria for progress in most scientific disciplines.

Progress in science means that the theories and laws become "better."

This implies higher accuracy of prediction and explanation, wider scope, increase of simplicity, and esthetic appeal. "Progress" further means that the trustworthiness of the leading system of paradigms increases. Indeed, scientific progress is not so much related to "scientific revolutions" but to the increasing consolidation of a given set of paradigms.

As a first approximation scientific progress may also be measured by the augmentation of the stock of objective data ("facts"). This implies, however, not only that the number of absolutely trustworthy singular propositions ("facts") increases, but also that the "facts" become organized and available for any actual need, be it theoretical or practical.

An estimate of the *future progress* ("potential," "fruitfulness") of a scientific discipline is difficult to achieve. Personal judgements of leading scientists and administrators of the granting agencies play a major part even if the reviewing process is performed by panels. As a rule, there is no straightforward choice of a research program in the competition with others that could be called "fully rational." It is in part at least a matter of science *politics*, rather than a scientific matter; and science politics are only slightly more rational than general politics. This is by no means a statement of principal criticism. I cannot think of any better way to administer science than it is actually being done by the leading agencies. However, as scientists we must be sober enough to realize that the demand for rationality generally accepted *within* the sciences cannot be met in science politics and science planning. Moreover, the public and the legislature will inevitably refer to "usefulness" when the further support of a scientific field is being considered.

There are a number of obvious obstacles to progress in the sciences that have their root in human nature. MAX PLANCK once said: "It is not that old theories are disproved: it is just that their supporters die out" (9). As every caricature, this statement contains the usual grain of truth. However, as I pointed out, some obedience or even allegiance to paradigms is essential, not only for the sake of continuity (tradition) in science, but also for the actual maintenance of stability in science as a largely intellectual enterprise. The paradigms determine the intellectual climate. This implies (at least for most of us) that they determine the manner in which we look at things in science and that they prevent us from seeing or believing things or features of things not in agreement with ruling paradigms. As BONDI puts it: "Sometimes attitudes that are possibly underlying particular points in theories and outlook are referred to among the cognoscenti as 'folklore'. There is a good deal of such folklore in physics, and in our interpretation and understanding of physics: a great deal of mythology even . . ." (9).

It requires a very strong, independent mind to advance a *new* conjecture or theory and to face and stand the attacks that will inevitably follow. It is to a large part a question of temper and natural boldness (and not predominantly a matter of intellectual brilliance) that determines whether a scientist prefers to work within the framework of established paradigms or dares to question and even to overthrow the venerable paradigms he grew up with.

The present state in elementary particle physics may serve as an example. As HEISENBERG recently pointed out (10), "many experimental physicists nowadays look for "quark" particles, particles with a charge of one-third or two-thirds of the charge of the proton. I am convinced that the intense search for quarks is caused by the conscious or unconscious hope to find the really elementary particles, the ultimate units of matter. But even if quarks could be found, for all we know they could again be divided into two quarks and one anti-quark, etc., and thus they would not be more elementary than a proton. You see how extremely difficult it is to get away from an old tradition."

Those who act as leaders of a scientific community must encourage the unconventional approach to a problem and be willing, e.g., as editors, administrators, or advisors in granting agencies, to invest risk capital. In most instances the usual peer review process for grant applications will favor—mostly unconsciously—the conventional and average approach. Too often creative minds become inhibited in their thinking and approach to fundamental problems by the criticisms of their less creative peers. Enthusiasm and devotion will deteriorate. The problem for the leaders of a scientific community is to recognize and defend the potential and fruitfulness of a new approach much ahead of the applause of the majority. This not only requires a sharp, independent, and imaginative mind, but also personal courage—a rare combination. On the other hand, the danger in investing too much risk capital may not be overlooked. The major problem is to maintain *a balance* between the support of well-established, high-quality endeavors and the support of new untried investigators and institutions.

I do not know of a system that would be better in principle than the present peer review process, "an institutionalized method of evaluating the quality, potential value, and feasibility of research proposals submitted to a funding agency" (11). "The peer review system . . . has deep roots in the structures and procedures of the scientific and scholarly community" (11). However, some obvious shortcomings and signs of corruption (or lack of fairness) could be eliminated. The shaken confidence of the public and the politician in the peer review system must be restored. This is not only a matter of administrative reforms. The question is *primarily* to what extent fairness, responsibility, wisdom, and a broadly educated mind are permitted to play the major part in evaluating the proposals advanced by the individual scientist. But ultimately, getting the *right* projects is fundamentally a choice between projects on the basis of promised quality. Under no circumstances should it be a question of responsiveness to political issues or pressures.

In recent months (fall' 75), the process of peer review of scientific projects has been the target of severe criticism in the USA. Some major questions—as summarized by MACLANE (12)—have come up: Is peer review fair? Does it provide for the support of the best science? Can it recognize potential breakthroughs? Are the reviewers chosen well? Do they respond objectively? As MACLANE points out, "such questions as these do not have any simple answer because they really refer not just to the immediate issue of how peer review works, but to the structure and nature of science in general and in particular. On closer examination, there are many different versions of peer review, each adjusted to apply to the science at issue. All of these versions have one purpose: to help decide how the limited funds available for the support of science can best be spent to advance both science itself and the national purposes to which it contributes. Science is a complex and multifaceted attack on the unknown. Guesses as to how it will work can most accurately be made by people who have themselves succeeded in such attacks on the unknown. These people are the peers" (12).

The peer review is largely designed to maintain quality of research. It is not the appropriate place for determining *political* priorities. Decisions concerning national priorities must be made by a political authority—on the basis of "if-then statements" supplied by the governance of the federal granting agencies or the leading academies *and* on the basis of national goals that transcend the realm of the scientific communities, e.g., *geographical* distribution of research support within the nation or actual needs on the level of "mission-oriented" research. However, once money *is* allocated to a particular field of research, the peer review system seems to be the most appropriate means to determine which projects within the field should be supported.

Progress is inevitably accompanied by some regress. The progress of science is seriously threatened by several regress phenomena, at least two of which are obvious and interrelated: "big science" and "scientific mass society." I want to describe these phenomena in the words of two bitter critics who know intimately the merits and shortcomings of modern science.

> CHARGAFF on "scientific mass society" (7):
> In those prehistoric times (BWC) [i.e., Before Watson-Crick, the creators of the central dogma; author's note] . . . science . . . was small; it was cheap; it was open. One could still do experiments in the old fashioned sense of the word. Now, everybody is working away at 'projects' the outcome of which must be known in advance, since otherwise the inordinate financial investment could not be justified. . . . The small numbers of scientific workers engaged in research had other consequences. It was easy to open new fields and to go on cultivating them; there was no fear of immediate dispossession as is bound to happen now. There were relatively few symposia,

and those that existed were not attended almost exclusively by hungry locusts yearning for fields to invade. Bibliographies were comparatively honest, whereas now entire packages of references are being lifted by a form of transduction, as it were, from one paper to the next; so that if some work gets into the habit of not being quoted, it never will be so again. The break in the continuity of the tradition has, perhaps, been one of the most disastrous effects of the scientific mass society in which we are living now. . . .

Another negative side effect of the scientific mass society will probably have immediate, serious consequences, since it is an important factor in widening the gap between the "Two Cultures" (13): overspecialization and concomitant underrating of the broad philosophical mind. This tendency in the modern scientific mass society is seen not only by concerned senior scientists such as WEISSKOPF (14); it is recognized in particular by those intellectuals who watch the scientific community from outside. SCHOLEM, president of the Israel Academy of Sciences and Humanities, bluntly states in a discussion about the "population explosion" within the scientific communities: "I maintain that the number of scientists who are aware of the implications and fundamentals of what they are doing has not grown very much . . . I have often wondered at how little people who had studied science,—chemistry, physics or biology—, knew about fundamental questions which should have been foremost in their minds" (15). And WEISSKOPF admits: " . . . most of the so-called scientists are not really scientists in a true sense, but they do constitute the majority of the scientifically trained. And in the science establishment the philosophical idea, the mystery of things, has become lost" (14).

It could be that the present cutting craze in America and Western Europe will turn out to be useful. However, the pruning of the tree of science must be done with wisdom and courage. The most important point is to cut out the dead wood—mediocrity.

HAGSTROM on "big science" (16):
Like other professions . . . modern science is characterized by the splitting of the professional role into the roles of the administrator and the technician. Leaders necessarily become politicised, oriented to obtaining funds and access to facilities and coordinating the efforts of others. Technicians become means-oriented, interested in performing their specialized skills for extrinsic rewards and uninterested in the recognition given by the scientific community for the attainment of scientific goals.

If this occurs . . . control is exercised by hierarchical authority *within* research groups and by political power outside of them. Scientists become more interested in their particular organization and in the reactions of politically powerful leaders than in the opinions of the wider scientific community. . . . When recognition from the scientific community loses its value, recognition will be sought from other users of research. As the norms of independence for professional scientists become abridged, scientists will come to feel less responsible for achieving scientific goals.

10. Progress in science has always been threatened by resignation. It happens from time to time that scientists who have achieved a strong position in their discipline are willing to say that the end of their discipline is in sight. Rutherford said at the turn of the century that atomic science is all over, and Max Planck proved that Rutherford was wrong. GUNTHER STENT recently said [In: *The Coming of the Golden Age: A View of the End of Progress* (17)] that biology has gone from the period of charismatic innovation and is now in the dogmatic and academic period. At least those biologists who work in fascinating fields such as neurophysiology or developmental genetics usually do not share this opinion. However, many of them show signs of concern about the predicted gloom of their science. It is generally felt that a crisis of progress in science will inevitably lead to a crisis of self-confidence among scientists. Self-confidence includes audacity and rigor; it implies that we trust our capacity to conjecture and to improve theories about the world in which we live; theories that deserve to be called consistent, coherent, comprehensive, beautiful, and deeply satisfactory; theories that please the creative human mind.

The scientific enterprise is fragile and vulnerable. With regard to what happened at the turn of our era in Alexandria, TOULMIN writes: "Fragmented into a diversity of subsciences, confused in the public mind with technology and craft know-how, divorced from the broader questions of natural philosophy, which had been the source of its original impulse and interest to educated men at large, the clear spring of the scientific debate inaugurated in Ionia and Athens disappeared into the Egyptian sands" (18). TOULMIN leaves no doubt that he describes a situation that was in some respect not unlike our own. He continues: "When I look at the fragmentation of the scientific enterprise today, at its divorce from any broader concern with natural philosophy, and at the wide-spread confusions of attitude that one finds both inside and outside the profession, as a result of which science has become so firmly linked in men's minds to technology—rather than to its traditional allies, such as cosmology and theology—I find the threat of Alexandrianism never far from my mind. While Copernicus was, undoubtedly, the man who initiated the *rise* of modern science, we may in our own day be in danger of witnessing the beginning of its fall" (18).

The feelings of insecurity and threat within the scientific community are being exploited ruthlessly by the press and other mass media for ideologic or simply commercial purposes. I think SAMUEL is right when he says: "This feeling of impending doom has been taken up by the press and other media, presumably because doom sells better than destiny, and the gloomy view of life is more newsworthy than an optimistic approach" (19).

Appendix

A Case History: Plant Morphology as a Mature Discipline Ruled by Paradigms

Fields that have come to a close, i.e., have reached full maturity, include classical thermodynamics of reversible processes or optical atomic spectra in physics. In biology most fields are still much less consolidated; they are still wide open for progress in the sense of normal science but also open for revolutionary progress. Some biological fields, however, have indeed reached a high degree of maturity, which makes dramatic changes of paradigms improbable, among them plant morphology. The paradigmatic state of this discipline has been challenged recently by SATTLER (20, 21).

In plant morphology the ruling function of paradigms, both in methods and techniques and in constructs and models, is particularly impressive, even for the dispassionate observer. The construct "leaf," e.g. (later on referred to as "phyllome"), has been of fundamental influence (in essence comparable to a platonic idea) in the thinking of many generations of plant morphologists since Goethe. The construct "leaf" permits the interpretation of all lateral appendages of a plant such as seed leaves (including cotyledons and coleoptiles), foliage leaves, sepals, petals, and stamens as manifestations of the same organ. The construct "plant" is based on a particular idea of plant construction, namely that the sporophyte of every higher vascular plant consists of only three types of organs: root, caulome (stem), and phyllome (leaf). All observable organs are manifestations of one of the three "subconstructs" of the construct "plant." The fruitfulness of these constructs can hardly be overestimated. The construct "plant" became a paradigm that seems to be resistant against any scientific revolution. As an example, the construct has survived the rise of the theory of evolution.

Evolutionary morphologists who claim that the higher plants with roots, stems, and leaves evolved from plants that lack these organs do not deny the eminent usefulness of the constructs "plant" or "phyllome." However, they reject any ontological or metaphysical interpretations of these constructs [as an example they reject TROLL'S essentialism in plant morphology (22)]. Homologization within a group of organisms that can be subject to the construct "plant" (in the foregoing meaning) does not imply the concept of evolution, but it has been used by evolutionary morphologists as a basis for inference on phylogeny. For fossil lower vascular plants, the construct "telome" [as originally invented by ZIMMERMANN (23)] is now generally used by evolutionary morphologists. They assume, of course, that a *continuous* evolution has occurred from lower to higher vascular plants. However, they also realize that distinct constructs are indispensable for organizing intellectually the *real* world. At least some of these constructs are deeply rooted and rule undisputably as paradigms or *Leitideen* the thinking of the whole field.

During the nineteenth and twentieth centuries the construct "plant," in particular the shoot construction (caulome, phyllome, adventitious roots), has been challenged by alternative models, but none of them has been accepted by the scientific community. At least for the angiosperms the classical construct has completely dominated morphological thinking. To what extent a new, interesting conception of the shoot, advanced in 1974 by SATTLER (24), will be accepted by the scientific community, remains to be seen.

At present the field of comparative plant morphology (which is, by the way, of considerable practical importance in agriculture, horticulture, breeding, and physiology) is concerned mainly with esoteric and highly sophisticated "puzzle-solving" within the framework of accepted paradigms. Is the so-called phylloclade of the Asparageae homologous to a caulome or mainly to a phyllome? Is the flower of Potamogeton, Triglochin, the Cyperaceae, Centrolepidaceae, etc., a "true flower" or an inflorescence? Is the stamen a phyllome or a caulome? Is the carpel a phyllome or a compound structure consisting of a fertile caulome fused to a sterile phyllome? Are the basic morphological units of the floral appendages phyllomes, gonophylls, or telome aggregates? This is the type of question [adopted from SATTLER (20)] being asked at present in the field of plant morphology. There is no reason to lament this state of affairs in terms of "lack of progress." Rather, this particular state of the field must be regarded as a clear sign of the *maturity* of a "classical" field of science. It is understandable, on the other hand, that a small number of brave young men rebel and sometimes rehearse a scientific revolution. SATTLER states ". . . that most of the interpretations in comparative morphology are 'certain or nearly certain' and only a relatively small number is doubtful. Now one might argue that a minority of doubtful interpretations is of little concern to the validity of the general theory which is supported by the majority of certain or nearly certain interpretations. But the history of science shows clearly that often it was the doubtful, exceptional cases which when taken seriously, showed the insufficiency of general theories and set a starting point toward the development of better, more comprehensive theories" (20). This is precisely the legitimate tension between obedience and loyalty toward the trustworthy paradigms, i.e., toward the intellectual achievements of our scientific predecessors and the feeling that existing paradigms do not suffice.

11th Lecture

The Ethics of Science

The discourse of truth is the ideology of science. We have defined science as a systematic attempt of the human mind to obtain genuine knowledge. Moreover, we have emphasized that scientific knowledge in a strict sense is "public knowledge." It is knowledge shared by the competent members of a scientific community. In this lecture I want to investigate the normative prerequisites, the intrinsic values of science, which enable the scientific communities to obtain genuine objective knowledge. As pointed out in the previous lecture, at least the intrinsic values of science must be shared by all members of the global scientific community. Otherwise science would not be a global cultural force.

The activity of every scientist is characterized by a strong social control. He is subject to a normative code which itself is based on the intrinsic values of science. Why does the scientist accept this normative code as compulsory? In HAGSTROM'S words (1): "Because scientists desire recognition they conform to the goals and norms of the scientific community." I believe that there are other motivations as well: because scientists *desire* to contribute to the growth and preservation of genuine knowledge they conform to the goals and norms of the scientific community. However, I must admit that scientists rarely want to be anonymous authors. Rather, they want to be known and they want to be recognized. HAGSTROM quotes a student who said: "Mathematicians want their work appreciated by others, especially by other mathematicians. You don't generally do things which won't be appreciated. One rule for usefulness is to get results which are useful in terms of other mathematical work" (1). As pointed out in the lecture on motivation, the striving for professional recognition is indeed probably the major motivation for scientists to conform to the normative code, at least it is a very reliable and steady motivation.

The normative code is a heterogenous complex. It consists of at least two parts: basic assumptions that are rigorously shared by the members of the scientific community and actual commandments. Since usually the assumptions as well as the actual commandments are not written up explicitly, the following lists are possibly not complete in the eyes of some scientists.

Among the basic assumptions we may discriminate two sets:

1. There is a real world (negative version: the notion of solipsism is not acceptable); the real world is intelligible; formal logic (including mathematics) is valid, without any restriction, in the description of the real world; there is no break in causal continuity.

2. Freedom of thought and freedom of inquiry must be guaranteed (this does not necessarily imply freedom in the choice of any particular goal; it implies, however, that the *result* of a scientific inquiry may not be influenced by any factor *extrinsic* to science); genuine, objective knowledge is *good*, i.e., it is superior to ignorance under all circumstances (this implies that there is no code of forbidden knowledge). Complete freedom *to publish* cannot be assumed in reality. The industrial scientist as well as the scientist working on classified governmental projects must be aware of the possibility that he may not be permitted by his employer to publish the results of his research without delay.

A tentative list of the actual commandments includes:

Be honest!
Never manipulate data!
Be precise!
Be fair (e.g., with regard to priority of data and ideas)!
Be without bias (e.g., with regard to data and ideas of your rival)!
Do not make compromises, try to *solve* a problem!

This list of fundamental commandments could be extended by more explicit formulations:

Use words and symbols with explicitly defined meaning!
Try to improve singular and general propositions with regard to inner perfection and degree of credibility!
Make accurate predictions and indicate the range of error!
Consider observational (experimental) data as the ultimate court of appeal!
Be ready, any time, to modify or replace a theory in view of inner inconsistencies or experimental refutation!
Always keep in mind that the members of the scientific community must depend on each other for reliability of material and intellectual methods, data, conclusions, and theories!
Regard *simplicity* as a high value! Do not create new constructs if not really unavoidable!

It is expected by the scientific community that the normative code is obeyed as a matter of course whenever a person works as a scientist, i.e., whenever this person is concerned with obtaining or processing data. The scientific community will not hesitate to respond to moral slovenliness with severe penalties. This ranges from withholding the recognition to expelling an individual from the scientific community.

The commandment "be fair!" requires a special comment since it is generally felt among scientists that this maxim is particularly difficult to comply with. Even great scientists have sometimes been charged with negligence. In her book, *Rosalind Franklin and DNA* (2) ANNE SAYRE claims that "Rosalind has been robbed, little by little" and she accuses scientific heroes such as Pauling or Watson of having been involved in this robbery. Since the DNA story is a very controversal matter (3), I have chosen a more clear-cut case to illustrate the obvious misbehavior of a great scientist. The following anecdote about Isaac Newton is particularly interesting, since it also implicates the problem of a moratorium in science, As PEKERIS tells the story (4):

> In 1776 Newton wrote to the secretary of the Royal Society, Oldenburg, recommending that a moratorium be put on the researches which were being conducted by Boyle, the author of Boyle's Law. Newton was particularly infuriated by an article which appeared a month before in the Philosophical Transactions of the Royal Society by Boyle entitled The Incalescence of Quicksilver with Gold. . . . This is a subject in alchemy, and what Boyle published in 1776 was a new recipe for making mercury react with gold and develop heat, and that is 'incalescence'. Newton, in one of the very few cases where he referred to other people's work, took the trouble to write to the Secretary of the Royal Society, asking him to intervene with Boyle to put a moratorium on researches on incalescence of quicksilver with gold because of the great menace such research would hold for the future of society. Now, closer examination of the motives that prompted Newton to write this letter turned out to be not so altruistic. Actually Newton was afraid that Boyle had the jump on him in discovering the philosopher's stone, as he, Newton, was heavily engaged in alchemical research. So people who are now proposing a moratorium on research should not in any sense feel they are suggesting something new.

We may conclude that irrespective of the practical difficulties with human nature, the "be fair!" commandment will be considered by every scientist as an indispensable maxim of the scientific enterprise. We expect as a matter of course our fellow scientists to be fair *at least* to the extent we comply with this maxim ourselves.

In order to avoid any misunderstanding I want to emphasize at this point that the cooperation among scientists, which includes mutual control and interpersonal criticism, is never free from severe tensions. Popper (5) has described the prevailing state of affairs as "the friendly hostile cooperation of scientists which is partly based on competition and partly on the common aim to get nearer to the truth." Scientists as a rule are not exceptionally ethical or modest in their *personal* lives and there is no reason to expect them to behave better than or differently from other people. Corruption, in particular nepotism, is not uncommon among scientists. Some scientists will even behave badly toward their fellow scientists as soon as competition, priority, prestige, and sometimes even money comes into play. Sometimes scientists dislike each other intensely. Even though these defects are not liked, they will eventually be ignored or tolerated by the scientific communities, if the scientist in question does excellent work and has never violated his real reputation, i.e., for being absolutely honest and trustworthy in *scientific matters*, that is, in obtaining and handling data and in using logic. Thus, a scientist does not have to fit into the same value system in his work and in his life, but if he does not he must be content to exist as a dual personality. Fortunately, most of us belong to this category of people. I strongly believe that for the scientist goodness, beauty, perfection, order, mystery, and God are not "counter-values" nor does he have to remove love and fear, awe and humility, humor and irony, admiration and hate, triumph and despair, tenderness and compassion from his emotional repertoire.

Certainly, many scientists are shy about *expressing* their emotions, but this may not be conceived of as an indication that their feelings have suffered because they are scientists. The occasional peak-experience or thrill a scientist is exposed to in the discovery of a new piece of truth does not reduce his emotional capacity, but he may become too bashful to talk about it publicly.

There is also an informal list of boundary conditions ("requirements for membership" in a scientific community) that should be fulfilled, more or less, if a person seriously envisages a career in science. The following points of this list deserve the attention of the novice:

1. I.Q. *and* creativity should be considerably above average. We recall in this context that I.Q. and creativity are not necessarily strongly correlated. While a feature of intelligence is the rapid recognition of complicated systems and the ability of equally rapid teleological action within the framework of a given system, unconventional thinking and tamed imagination are major attributes of creativity.

2. The scientist needs a strong and *persistent* motivation. Scientific work usually requires considerable effort and a strong discipline of thought over longer periods of time. Persistence, but not stubbornness, is a very important prerequisite of successful scientific work.

3. A scientist should maintain some features of the intense curiosity of a child—although in a cultivated and disciplined form; however, sometimes a modest *disinterest* in people and social relationship will be indispensable for the solution of a problem that requires the full concentration of the human mind. Scientific work always implies some renunciation.

4. Some sense for conformity is indispensable. We recall that science is a social phenomenon, a matter of communities. Therefore, a minimum of conformity with the rest of a scientific community is a prerequisite for membership.

Returning to the structure and to the value system of the scientific community I want to consider three aspects in some detail: the notion of agreement, the consensus principle, and the value of freedom. In the forthcoming lecture I will investigate the possible reasons for the present crisis of science as a cultural force.

Agreement in science is difficult to achieve since scientific communities are composed of unusually individualistic and ambitious people. On the other hand, *some* agreement is obviously essential for the scientific endeavor (see definition for science on p. 30). We consider the problem of "agreement" on three distinct levels.

1. Agreement with regard to basic judgements (singular propositions) is essential for science. The notion of "fact" or "objective data" is a prerequisite of science.

2. On the theoretical level (theoretical propositions) perfect unanimity can hardly be obtained and is in fact not even desirable. Nearly perfect agreement on points of theoretical interest is characteristic for mature theories, relatively closed fields, ruled by strong paradigms. Newtonian mechanics, classical thermodynamics, comparative morphology, or the field of optical atomic spectra are examples.

3. With regard to intellectual approaches and material methods, *pluralism* must be encouraged! On the other hand, those methodological approaches that have been proved very successful in obtaining and processing data must be preserved and transmitted to the next generation. Since we have symbolic language and print at our disposal, tradition of this kind must not necessarily be related to a ritual. However, there are methodological rules that deserve deep respect and even admiration. It is part of sound tradition in science to teach the beginner a number of methodological rules quite authoritatively. At the same time, however, the student must be assured that a trust in methodological tradition and authority is compatible with open-mindedness. Actually, the scientist must remain open all his life for new ways to look at a problem even though with increasing age preference will be given to those methodological and logical approaches that have proved successful in the past. This tendency can become dangerous. Sometimes methodological inflexibility of influential scientists will lead to hopeless degeneration of a research program.

The notion of *agreement* in science implies that we must *trust each other*. This strong belief in and reliance on the integrity of our fellow scientists is not only basically important with regard to data but also with regard to conclusions. For example, no author of a scientific textbook can base his writing exclusively on the study of original papers. He depends heavily on review articles and on monographs. Thus, the intellectual organization of scientific knowledge involves several steps that depend on each other. The integrity and trustworthiness of our peers is an indispensable element in every attempt to codify genuine knowledge. Even though trustworthiness of our peers may be taken for granted, the writing of a textbook or a monograph is always a risk. Small errors can accumulate and can amount to large hazards, especially when the work is destined to be used both by scholars and students *and* by policy-makers as a major source of established knowledge on its subject. Furthermore, the reader of a monograph may be tempted to attribute greater certainty to conclusions than was intended by the author if the conclusions agree with his prejudice. While it is unfair to blame writers for the abuse of their work by readers, writers must be aware of this danger and draw their conclusions in view of this risk. The peer review system for scientific books is an extremely valuable control system if handled rightly, i.e., without bias and with a maximum of fairness and competence, in brief: in accordance with the normative code of science.

The *consensus* principle (6) formulates the "ideology" of agreement in science. It implies that, in order to become *established* scientific knowledge *in a strict sense*, a singular or general proposition must be accepted by most of the competent people in a given scientific community. The consensus principle has obvious disadvantages since it favors the average scientist who strongly adheres to established paradigms at the cost of the creative or ingenious member of a scientific community. However, there seems to be no escape: in order to maintain the notion of objectivity (intersubjectivity) we must be prepared to cut back the wings of those whose creativity, intuition, or methodologic skill is beyond the grasp even of the leading people within a scientific community at a given time. This is a bitter procedure for some individuals, but even a partial sacrifice of the consensus principle would cause the scientific communities to fall apart. We must always be aware of the possibility that a (subconscious) prejudice or an ideologic bias can affect the data and/or the conclusions a researcher draws from his or from others' data. There is no principal cure except the consensus principle. Thus, the concensus principle must be maintained even though this principle can easily be blamed *in retrospect* to have in fact inhibited the progress of science. HEISENBERG tells the following anecdote (7): " . . . when I entered Niels Bohr's institute in Copenhagen in 1924, the first thing Bohr demanded was that I should read the book of Gibbs on thermodynamics. He added that Gibbs had been the only physicist who had really understood thermodynamics." Has it been fair to force Gibbs under the rule of the consensus principle?

The negative side effects of the consensus principle have become more obvious with the rapid quantitative growth of the scientific communities. According to KATZIR-KATCHALSKY (8), Derek Price, an American statistician of science, has found that if the number of scientists is n, the number of scientists producing fundamental results is $\sqrt{n}$; hence the ratio of $\sqrt{n}$ to n *decreases* with increasing n. Also from the point of view of allegiance toward the consensus principle, the dilution of superior talents during the recent numerical growth of the scientific communities must be considered a very serious regressive phenomenon.

This antagonism between the superior (sometimes young) scientist and the scientific establishment within a given scientific community is a heavy burden for the peer review system of scientific journals. Often an enthusiastic author will get the impression that the referees of submitted manuscripts consider the suppression of progress as their main task. It is easier, indeed, to get a paper in press with solid but hardly exciting data ("On the influence of cyanide on seed germination") than a paper with data or ideas that burst the framework of established thinking. It requires the superior knowledge and wisdom of an experienced, but nevertheless flexible editor, to maintain the right balance between sound scepticism against the new and unexpected data and ideas (which might be wrong and misleading altogether) and the optimistic belief in the methodological, intellectual, and moral abilities of a superior fellow scientist.

The consensus principle can, of course, be deteriorated by unwise use. A recent example is a vote on a draft resolution on Heredity, Race, and I.Q. (9). This resolution was circulated by the Genetics Society of America (GSA) to its membership, because the GSA governance had decided to poll its members before publishing any resolution in the society's journal, *Genetics*. The poll was unwise for at least four reasons:

1. Scientific truth or falsehood is not decided by vote. This modus procedendi is a caricature of the process that usually leads to consensus within a scientific community of *competent* people.
2. The 796-word draft resolution covers a broad spectrum of scientific, social, and political concerns: from the efficacy of I.Q. testing to the support of research in education, from the cultural and environmental differences between blacks and whites to the duty of geneticists to oppose racism in the academic sphere. It would have been much wiser to separate carefully the scientific statements from those that clearly transcend the realm of scientific statements on this matter.
3. The major shortcoming, competence, has been expressed by a concerned GSA member (10) as follows:

> . . . it must be recognized that the GSA membership, being human, will reflect in their responses to this statement an unknowable degree of ignorance of the facts and of political and cultural bias. Genetics is such an enormous, all-embracing subject, including as it does the study of biochemistry, cellular physiology, tissue development, plant and animal breeding, population genetics and evolution, that many geneticists are no more familiar with the statistical treatment of quantitative genetic traits than are scientists in very different disciplines . . . the response of the GSA membership as a whole is trustworthy only insofar as we can rely on many members' willingness to admit that they are insufficiently informed.

Unfortunately, the latter presupposition was far too optimistic. Only a very small percentage of those responding elected to disqualify themselves on this basis. Thus, the question remains as to the qualifications of those who chose to agree or disagree. In fact, it is only a relatively small number of people who are really familiar with this difficult matter, including predominantly anthropologists, psychologists, and (more recently) developmental (human) geneticists. The bulk of the GSA is clearly not qualified to operate as a decision-making body in this difficult matter, which requires highly specialized expertise. Fortunately, a most comprehensive, critical, and balanced review of the race–I.Q. issue has recently appeared (11). This book, meticulously written by three cautious, qualified scholars (professionals in psychology and anthropology), should help the GSA to withdraw from this matter and to avoid a further fall into disrepute.

4. The resolution is poorly phrased. For example, it carries the statement: "There is no convincing evidence of genetic differences in intelligence between races" (12). A fair scientific statement about this matter would be *symmetrical*, stating that the presently available scientific information neither confirms *nor excludes* genetic differences between human populations with regard to the distribution function of I.Q. scores.

Freedom of inquiry is an indispensable prerequisite of the scientific enterprise. Any extrinsic constraint is not acceptable. On the other hand, the *intrinsic* constraint, the obedience to the ethics of science, is compulsory. I have already mentioned that freedom of thought and freedom of inquiry must not *necessarily* imply freedom of choice of any particular goal. Society or any private employer of scientists must have a right to set certain goals. The right of society or of some industrial employer of scientists to demand that certain problems, which are practically relevant, have a high priority in research is legitimate and can hardly be denied. Freedom of inquiry implies, however, that the result of a scientific inquiry may not be influenced by any factor extrinsic to science. Another implication is that, *potentially*, every goal can be chosen. Any "index of forbidden knowledge" or moratorium is not acceptable to the responsible scientist.

The catastrophic consequences of constraining freedom of inquiry for *ideological* reasons are well illustrated by the cases of Galileo (the loss of Italy's leading position in the rise of science during the Renaissance) and Lysenko [the loss of genetic knowledge and concomitantly of agricultural potential and productivity in the Soviet Union during the Lysenko–Stalin era (13)].

A recent target for a moratorium has been the research on the genetic vs. environmental contributions to human intelligence. Besides the fact that also in this field, ignorance and prejudice are hardly better than objective knowledge, it is a serious sign of immaturity if even some scientists insist that this particular line of investigation be forbidden or at least discouraged. The scientific enterprise cannot be divided up. To put a moratorium on a particular line of research means that *every* field of research is potentially subject to a moratorium. There will always be some who maintain that knowledge in some areas is undesirable because it might destroy myths, beliefs, or convictions regarded as indispensable. In the past this kind of objection against scientific truth used to be based on religious beliefs; now it is mainly political or ideological prejudice, in particular the awkward ideology of egalitarianism. "Everyone committed to science must utterly reject and oppose the doctrine that ignorance is better than knowledge, self-deception better than intellectual honesty, faith better than thought" (14). Otherwise every piece of scientific enterprise would be subject to prejudice at any time. I stated previously that any index of forbidden knowledge cannot be accepted by the scientific community because it is not consistent with the ethics of science. This implies that any kind of *censorship* cannot be tolerated by the scientific community.

The fact that a problem may prove politically embarrassing is no reason for invoking constraint. Although knowledge alone does not guarantee success, its lack almost certainly reduces the chance and extent of progress. Politically embarrassing problems are precisely those where objective knowledge is required the most. The technological *application* (in a wider sense) of genuine knowledge requires additional decisions that are beyond the realm of science. A decision by the decision-makers of society may forbid the application of knowledge *in practice*, but the *acquisition* of knowledge must remain free. Freedom of inquiry implies, however, being subject to the ethics of science. The normative code of the scientific community is in fact the only constraint for the scientific enterprise.

This constraint becomes effective if certain experiments turn out to be dangerous for human life. It is clearly part of the ethics of science to stop an experiment whenever death or severe injury of a person comes into play.

SAMUEL (15) has suggested "a code to guide scientists working on the brain. It should deal with the ethics of the use of human volunteers for experiments including laboratory investigations of social and psychological pressures, which can be as damaging to the brain as drugs, with the permissible extent of trials of chemicals that affect the personality; and with the moral aspect of transplanting parts of the brain and of reviving patients with severe brain damage." This code, not unlike the Hippocratic Oath, could indeed become an ethical postulate for all those scientists concerned with the function of the brain. As SAMUEL continues: "This code . . . should help us, as the secrets of the brain are steadily unravelled, to preserve the sanctity, the integrity and the individuality of men's minds." Irrespective of our particular interests we probably all agree that concern for human dignity puts some limits on research. However, the ambivalence of the case may not be overlooked. As DAVIS (16) remarked: "I am worried that the strength of our reactions to sins of commission, in the testing of drugs and other new therapeutic devices, may bring to a standstill further advances in therapy. No new drug has ever been produced that hasn't involved risk for the first people and a great many drugs have been dropped after a great many people were harmed. But we wouldn't have medicine without somebody assuming those risks."

A related point is the preaching of extreme caution in handling potentially hazardous material, such as plasmids, radioactive material, antibiotics, and drugs. The so-called Asilomar Conference on recombinant DNA molecules in late February, 1975 (17,18) tried to deal with the potential biohazards of this work. The experimental technique in question involves the use of recently discovered enzymes to rearrange the genetic material of different organisms in novel combinations. This offers an unprecedented means of joining DNA from different organisms in combinations that may never before have occurred in nature. The potential benefits resulting from the considerate application of this technique are enormous: recombining DNA could lead to advances in medicine and agriculture in the long run and to new insights into basic genetic mechanisms, including developmental genetics, in the short run. However, there is a risk of undesired, hazardous side effects: the escape of organisms containing novel hybrid DNA might inadvertently cause the spread of infectious diseases, carcinogenic agents, or it might extend the range of antibiotic resistance.

Ending a six-month period of voluntary suspension of those experiments considered most dangerous by a distinguished committee of 11 scientists, the Asilomar Conference—attended by a select group of 140 from all over the world—gave the green light to resume the work on recombinant DNA, but only under the stringent safeguards established during the Conference (17). In the following months extraordinary difficulties have arisen in translating the general principles laid down at Asilomar into practical guidelines that everyone can live with. A. STETTEN, chairman of the NIH committee known as the Recombinant DNA Molecule Program Advisory Committee, put it: "We are being asked to set guidelines based upon hazards based upon accidents which have not yet happened. Even Lloyds of London is unwilling to write insurance on accidents for which there are no actual data" (19).

Since there seems to be agreement that a moratorium could and should not be maintained over a considerable period, the general feelings at present are probably best described by a statement of the British working committee under Lord Ashby, which recommended as early as December, 1974, that "subject to rigorous safeguards," recombinant DNA techniques "should continue to be used because of the great benefits to which they may lead" (20).

A major problem has been whether or not the guidelines should be so specified and rigid that some kinds of experiments that are considered particularly hazardous cannot be undertaken without violating the guidelines. This would mean a de facto moratorium on some experimental approaches—an awkward idea.

What is required in practice are safety guidelines for the various types of recombinant DNA experiments similar to the safety guidelines developed previously for the use of radioactive materials in biologic and medical research and practice.[7] The appropriate safety precautions standard must be determined by law, on a national and, if possible, on an international basis, the law being based on expert opinion *and* on the experience and common sense of the lawyers and the members of the legislatures. At all costs, the scientific community involved in this matter must not be allowed to become polarized into opposing camps, depending on their interest, or lack of interest, in experiments involving recombinant DNA. Tensions within the scientific community are already far developed. For example (19), a member of the Genetics and Society group of Scientists and Engineers for Social and Political Action, Jonathan King of the Massachusetts Institute of Technology, says that the function of the above-mentioned NIH committee, as presently constituted, "is to protect geneticists, not the public." Hogness, chairman of the subcommittee that wrote the original guidelines, is an active worker in the recombinant DNA field, which King likens to "having the chairman of General Motors write the specifications for safety belts." While the metaphor is poorly chosen (why should the chairman of General Motors not be interested in *optimum* safety belts?), the example illustrates that even within the scientific community the recombinant DNA problem has in part become a political matter.

To emphasize once again: the notion of a moratorium is awkward. This has been recognized by most people who have carefully thought about this problem. As Aharon Katzir-Katchalsky, who was a fabulous scientist and a great man, said (21):

> . . . nobody knows how to stop research, even if such a decision is made. And surely nobody knows how to keep truth under control and stop its dissemination. In Bertold Brecht's Galileo the Great Inquisitor pleads with Galileo and asks him to put a moratorium on his astronomical discoveries. The pleading of the Inquisitor has great human significance. He tells Galileo: 'With your little telescope you have torn down the skies and instead of skies you have now an empty space. I have no place left to put God and the Angels . . . And he says more: 'The central position of the globe which gave so much meaning to human life is undermined. Have pity on mankind and keep it quiet.' And the answer of Galieo, a cruel, horrible answer, is the classic answer of science; he has full sympathy with the Inquisitor, but you cannot keep truth in hiding. If he suppresses it, somebody else will disseminate it. . . . science is primarily the great venture of man, the venture of understanding Nature, the great spiritual adventure, and to stop science is essentially impossible.

I would add that we can stop science by destroying it, destroying its spirit, its tradition, its paradigms, its institutions, its freedom. Science is a

[7]An advisory committee of the National Institute of Health (NIH), which met at La Jolla on December 4th and 5th, 1975, has finally drafted a set of strict regulations that are likely to receive broad international consent (see Nature (Lond.) 258, 561, 1975).

result of cultural evolution, it is only quasi-stationary and will always be unstable unless it is permitted to grow. "Growth" does not necessarily imply a *quantitative* increase; rather it implies growth in quality, improvement, optimization. Qualitative "growth" (optimization) of science must be maintained as a permanent process in order to compensate for the inevitable regressive phenomena of society and technology. Society is always threatened by random failures and negative side effects of teleological action. It is the gain of scientific knowledge that can overcome the loss of neg-entropy in our cultural institutions.

No scientist, however, should be forced to do research he does not want to do for ethical reasons. If a scientist feels that society would inevitably and immediately use the knowledge originating from his research for purposes he considers to be *evil*, he must have the right to give up at any time. But the scientist may leave no doubt that his decision is based on his *personal* value system, which is *extrinsic* to science. An illustrative case from the late sixties is James Shapiro, who declared his intention of renouncing his scientific career at the AAAS meetings in Boston in 1969. He had been part of the Harvard team that isolated the lac-operon. As reported by SHINN (22), Shapiro argued "that so long as men like Nixon and Agnew prescribe what is going to be done with scientific knowledge, we scientists should quit giving them the materials for their bad decisions." Shapiro can hardly be blamed as long as he speaks as an individual. However, can we still follow him when he tries to make his conviction compulsory for the scientific community? Can we be convinced that he has thought enough about the relationship between the scientific community, society, and political power in different parts of the world? Or about the relationship between knowledge, wealth, security, and power?

It has not escaped my notice that for an industrial scientist, the situation can become very difficult when he has to decide between making available a piece of information (e.g., about a serious hazard to workers in a plant) *to the public* and loyalty toward his employer. Similar situations are well known from industrial medicine. For the neutral observers, e.g., the peers of the particular scientists, the dilemma is still more complicated because they must consider the possibility that the employee may be acting not in the public interest, but out of personal or ideological prejudice or out of spite against the employer.

12th Lecture

Science and Technology

Genuine scientific knowledge is the most important means of teleological action in the modern world (see Fig. 11). Teleological action is the essence of human culture. Is it fair to blame science for the *kind* of use, often an unwise and destructive use, of our knowledge in the context of teleological action? The philosopher TOULMIN (1) writes:

> . . . Other critics attribute current anxieties about the moral justification of science to its impact on the natural and human environment, and their views certainly appear to have a little more foundation. There may of course be some reason to foresee a possible future in which the applications of natural science to technology might have a drastic impact on the quality of human life, even on the possibility of its continuance. Yet, once again, there is an element of exaggeration in the current debate which distracts attention from the true sources of our present ambiguities. If we leave aside the special problems created by the use of non-degradable pesticides and similar chemical agents, the remaining basic sources of environmental pollution are nearly all of them the present day products of historical processes going back to the year 1800 or before—certainly, long before any serious application of scientific ideas to industry or manufacture. These sources are (in brief) population growth, urbanization, and ill-controlled industrialization, all of them factors whose chief origins lie outside the scientific movements. Thus, if we end up by having more children than we have the resources to support, that is not the fault of science. If cities and factories use the earth's air and waters as an open sewer, that is not the fault of science either. So it need not surprise us to find all the main themes of the contemporary ecology movement anticipated long ago, in the writings of men like Thomas Malthus, William Blake, and Anatole France. Even apart from the whole of 20th-century science, we would still be overloaded with Blake's 'dark, Satanic mills'; and without the help of contemporary science these factories might well have been even more damaging to the environment than the ones we actually have. Arguably, indeed, there seems to be too little application of

> scientific thought and analysis to the industrial organization and practice, not too much. Industrial technology is industrially damaging not because it is too scientific, but because it is too unscientific.

I have written previously (2):

> The goal of technology is to change the world according to decisions made by man. The data and laws and in particular the paradigms of science are the most important elements in modern technology. Our present world is totally adapted to and dependent on technology, and thus technology is an irreversible phenomenon. The negative side effects, the regressive phenomena of technology which threaten human life (and the life of every other creature as well) can only be overcome by the use and considered application of more and new technology which is again based on genuine scientific knowledge. Since many of these data are not yet available or not reliable enough to allow the formulation of laws, the progress of science is essential for our technological survival . . . It is not only 'mission-oriented' research which is urgently required (by this term I mean work done where the area of application is known); a similar emphasis must be applied to 'curiosity-oriented' research. By this term I mean research done exclusively for the furtherance of genuine knowledge and thereby strengthening the intellectual framework of science, the trustworthiness of the present day paradigms on which any piece of technological innovation is at least indirectly dependent. Therefore, it would be a disaster for our own and every future generation, if the rate of scientific progress were considerably reduced.

In considering the intrinsic value system of science, the ethics of science, I have defended the thesis that genuine knowledge is *good* in an ethical sense. In technology the situation is totally different. Every technological achievement is necessarily *ambivalent:* that is, it can be good or bad, depending on one's point of view or on the particular situation. Technology is *necessarily* a double-edged tool. The classification of a given technological achievement as good or bad, right or wrong, is never certain. Any given piece of technology can always be classified as ethically good or bad depending on the ends one has in mind, and depending on past, present, and future boundary conditions.

The nuclear bombing of Hiroshima, which was intended to be and conceived of as an ethical act (to save the lives, both American and Japanese, which would be lost in a full-scale invasion), was later classified as unethical.

Jet aircraft are often regarded—from the point of view of pollution and of the impact of science on society—as an antihuman device. On the other hand, as LIGHTHILL pointed out (3), the transformation of the world into a neighborhood by aircraft makes it possible to decrease the amount of prejudice and superstition among people and to extend the altruism that originated from our behavior and feelings in the family or tribe to the globe as a whole. Thus, this piece of technology is making possible man's greater humanity, because it makes it easier to love one's neighbor on the other side of the globe.

The transistor was invented in 1948 on the basis of scientific advances in the field of solid state physics. Within less than a decade this device not only revolutionized a good part of electronics and electronic computing, but also profoundly affected and changed the whole world, i.e., society as a whole. There is hardly any aspect in science, technology, medicine, industry, communication, politics, and the arts that has not been influenced deeply by the development of semiconductor physics and by the invention of the transistor. Who could dare to apply the notion of 'good' or 'bad' to the original research, to the invention, to the numerous innovations, and to the widely ramified technological applications of the original invention? The newer technologies such as those of computers, automation, and communications are often cited as principal causes of dehumanization. This is short-sighted and unfair. In our present context it can be considered as an instance when "technology is being made the scapegoat for failures in human ethics" (4).

The chemical DDT is another illustrative example. Its use brought a dramatic halt to a cholera epidemic in Naples in World War II, and its initial success in destroying agricultural pests was spectacular. However, the gradually appearing side effects, in particular the progressive concentration in food chains, has led in the United States and in West Germany to the banning of DDT for nearly all uses. On the other hand, DDT could not yet be replaced as an essential component of any major program of malaria control (5).

Most people are horrified by the perils of the uncontrolled use of drugs, in particular addictive drugs, and the wave of crime and extortion that follows in their wake, and at least some blame pharmacological chemistry for having created or at least made available in large amounts these terrible molecules. However, as SAMUEL pointed out (6), "we must not forget that through the use of drugs a way has been found, at last, to alleviate pain, to reduce anxiety, to end insomnia, to control violence in the mentally disturbed and to compensate, in part, for the loss of concentration in the ageing."

These examples may suffice. They stand for an infinite number. To repeat: any technological achievement can be classified as ethically good or bad, depending on the ends one has in mind, and depending on past, present, and future boundary conditions.

From the *point of view of ethics* the relationships between science and modern technology can be described by the following statements:

1. True scientific statements are good in an ethical sense.
2. Every true proposition, every piece of genuine knowledge, can potentially serve as a means in teleological action (see Fig. 11), i.e., can potentially be applied in technology.

3. Every technological achievement, including anthropo-created (man-made) ecosystems, is always and necessarily ambivalent. Therefore, every piece of human culture, which inevitably implies teleological action of man on nature, is always and necessarily ambivalent.

4. It is a grievous misunderstanding of science to interpret the inherent moral and factual ambivalence of technology as an ambivalence of genuine knowledge.

5. While the development of new and the maintenance of old technologies must be subject to political decisions, the progress of science must not be controlled by political power. Any "code of forbidden knowledge" cannot be accepted by the scientific community. On the other hand, the decision-makers of society can and must forbid a technological development if they are convinced that a particular goal is not worth the risk.

6. In a free society that is necessarily pluralistic you will always find a multiplicity of goals and a diversity of opinions concerning technology assessment, including risk evaluation. For this reason any political decision with regard to technology is necessarily a compromise that can never please everybody. Let us assume we would be able to do genetic engineering in man. Could we agree about the goal? What kind of human being should be preferred, intelligent and coolblooded, emotional and bold, creative and sensitive?–Clearly, the problem is that engineering requires goals, and goals depend on decisions that imply values (see Fig. 11).

What is the function of *science* in decision making, technology assessment, and risk evaluation? I outlined briefly in the Prologue a two-step model of decision making that implies a rigorous and consistent cooperation between scientists (including science-oriented technologists) and decision-makers (that is, political representatives of the people) wherein the responsibilities are clearly defined. In this model the decisions (including those about top research priorities) are to be made by the politicians, but they may only be made between alternative models that have been elaborated, or at least approved, by scientists familiar with and competent in the particular matter in question. The scientist's responsibility is to ensure that only genuine knowledge is considered and that the ethical code of science is obeyed during construction of the alternative models. On the other hand, the politician is responsible for the decision to be made between the different models. At this point, different value systems and propensity structures rightly come into play. Let me repeat briefly the practical procedure that must be followed if we apply the two-step model of decision-making in reality. Using different assumptions, different self-consistent systems can be constructed, each of which is equally logical and equally justified by scientific knowledge. The alternative models (or systems) constructed by the scientists are only different because different *assumptions* have been chosen. In most cases the solution of a complicated problem in the real world can be based on different assumptions depending on the means and values one has in mind (see Fig. 11), and a *decision* to use one assumption rather than another is required. In most cases the scientist cannot make the decision because several sets of assumptions are equally justified from the point of view of science. In those cases, a political decision based on *political* experience, on *political* taste, and on a *particular* value system and propensity structure ("ideology") come into play and are in fact indispensable. The decisive points in the two-step model of decision-making are that the responsibilities are clearly defined and that the scientists are prevented from usurping *political* leadership. As a rule, scientists move into the arena of public policy without having learned the finesse of politics. Most scientists are politically naive and inept at taking over political responsibility and handling political power. I do not know of any scientist who has possessed the attributes of statesmanship, and if anybody has ever been in an academic senate he does not want any more university professors running the world.

A serious problem has been the communication gap between the establishment of the scientific community and the legislative leaders (e.g., in the U.S. Congress). According to CURLIN (7), "Scientists have attributed the problems of the political community to the domination of politics by lawyers, who are allegedly trained to win cases rather than to solve problems. . . . An attitudinal survey was conducted in 1972 among members of the American Bar Association's Natural Resources Law Section, a group of lawyers who are in close contact with scientists and engineers. The results were startling. A significant number of the 575 respondents questioned the objectivity and veracity of scientists—qualities that are considered to be fundamental to science. Lawyers also perceived scientists to be narrow in their social outlook and provincial in their approach to problems."

Steps must be taken to improve interprofessional appreciation. It seems that close communication during the actual legislation process is the only means to remove the barriers of mutual misunderstanding. The maintenance (and recovery, if necessary) of public confidence in the honesty and objectivity of scientists must be a prime concern of the scientific community.

The views I have just outlined are by no means new or original. They were expressed by HAROLD LASKI (8) as early as 1931 with a somewhat different emphasis:

> It is one thing to urge the need for expert consultation at every stage in making policy; it is another thing, and a very different thing, to insist that the expert's judgement must be final. For special knowledge and the highly trained mind produce their own limitations which, in the realm of statesmanship, are of decisive importance. Expertise, it may be argued, sacrifices the insight of common sense to intensity of experience. It breeds an inability to accept new views from the very depth of its preoccupation with its own conclusions. It too often fails to see round its subject. It sees results out of perspective by making them the center of relevance to which all other

> results must be related. Too often, also it lacks humility; and this breeds in its possessors a failure in proportion which makes them fail to see the obvious which is before their very noses. It has also a certain caste-spirit about it, so that experts tend to neglect all evidence which does not come from those who belong to their own ranks. Above all, perhaps, and this most urgently where human problems are concerned, the expert fails to see that every judgement he makes not purely factual in nature brings with it a scheme of values which has no special validity about it. He tends to confuse the importance of his facts with the importance of what he proposes to do about them.

It is unthinkable in a democratic and thus inevitably pluralistic society that scientists could actually become endowed with the obligation and with the authority to assume full moral responsibility for the social impact of science. In a free, pluralistic society the experts' judgement can never be final. It is the statesman and the experienced politician, not the scientific expert, who is the decision-maker. The function of the expert is clearly restricted; he is the creator of the if–then propositions on which the responsible policy-maker should base his decisions.

Technocracy, i.e., the control of society by a scientific–technological elite, is no attractive alternative since it inevitably implies dictatorship. Dictatorship means that a single man or a few men determine, by crude physical or sophisticated psychological power, the valid value system as well as the official propensity structure, or, in simpler terms, determine what is good for the people.

I have already emphasized that under most circumstances decisions to be made by any legislature, government, or administration involve the experts if–then propositions as well as value judgements and that the latter nearly inevitably leads to divergences of opinion. However, often even the if–then propositions become a matter of controversy as soon as they involve *extrapolations* of genuine knowledge or currently available technology. This will usually lead to conflicting claims and sometimes even to serious controversies within the scientific-technological community. Inheritance of I.Q. and the risk of nuclear power plants are cases of this kind.

Moreover—as pointed out in the prologue—the moral responsibility a scientist feels as a citizen of a country or as a partisan of a particular political ideology can easily affect his judgement as to the state of scientific fact, in particular when the pertinent scientific data are not yet thoroughly "objective." Even benevolent critics of the scientific community hold the opinion that it is not possible for scientists "to have deeply held moral and political views about a question and simultaneously maintain complete objectivity concerning its scientific components" (8). This is a fair description of a sad situation that can only be mastered if at least the *leading* scientists of a scientific community abstain from political ideology and leave no doubt that responsibility toward "the people," public responsibility, lies clearly with the political representative whereas the scientist is primarily responsible for the truth of statements and thus for the integrity and credibility of the scientific community. The question to be decided by the scientist is not "should we build nuclear power plants; it is, rather, "can a 'safe' nuclear power plant be built" (whereby 'safe' must be operationally defined!). Another example: the question to be decided by the scientist is not "should smoking be prohibited by law" but, "is there a causal relationship between smoking and lung cancer or heart diseases or perinatal mortality." A final example: the question asked by the scientist is not "should a man who has a XYY sex chromosome constellation be left in complete freedom and remain unobserved until he commits some horrible crime," but "does a correlation really exist between criminality and XYY incidence in human males."

Of course, the scientist may have a very strong opinion about the desirability of a measure, a project, a law. With regard to *this* question, however, he is nothing more than a member of a democratic, pluralistic society (in the free world). With regard to *desirability* he is *not* an expert. As pointed out previously, technocracy is no attractive alternative as compared to the democratic process. From my own experience I agree with CHURCHILL'S aphorism that "democracy is the worst form of government except all the others." I trust that most scientists agree with the democratic state in essentials.

The worst problem arises if the views of different experts (e.g., "progressive" and "conservative" scientists with regard to their respective propensity structure) cannot be reconciled because objective information is still lacking. If an authoritative statement of a scientific advisory committee is nevertheless required because the practical problem to be tackled is urgent and vital, some kind of a "neutral instance," a scientific judge, must be found to evaluate the opposing or diverging views of the experts. Under all circumstances the impression must be avoided that in critical issues, opposing political groups will hire their own group of scientists to provide them with the "facts" required from the point of view of a particular political position (8). A neutral judge from within the scientific community must be found in these cases to prevent further spoilage of the image of the scientific advisor and damage to the prestige and integrity of the scientific community.

I want to analyze now the relationship between science and technology *apart* from the ethical implications.

The ascent of science in the seventeenth century was closely related to an improved craftmanship in making more accurate instruments, such as clocks, balances, or microscopes that permitted precise measurements and thus an exact testing of scientific theories (9). Science and craftmanship advanced hand in hand, and this has remained an essential aspect of the observational as well as experimental sciences. In the modern world science and technology can be looked at as a gigantic feedback system. Science is the basis of technology, and it is the progress of technology that supplies the observing or experimenting scientist with increasingly better means. The progress of *measurement* in science depends entirely on the progress of technology. In biology, the electron microscope and the ultracentrifuges, including reliable vacuum pumps, are particularly spectacular examples.

Often the same person will be both a scientist and a technologist. It depends on the particular objectives. I have worked for years as a "technologist" in an effort to improve the technology of generation and measurement of monochromatic light, although the final objective has always been to study the effect of light on plant development. Even from a psychological point of view the relationship between basic science and applied science (applied in the realm of technology) is close and complicated. Many good scientists are strongly interested in the practical application of their results even though they have clearly thought about the inevitable and unresolvable dilemma of "good" and "bad" in any piece of technology. Why? Apart from the financial aspect, the strong motivation of the scientist to apply his knowledge in the real world stems from the notion that the conspicuous practical success of a discovery convincingly documents the fact that we have really "understood" a piece of nature. HEISENBERG (10) reports the following anecdote: "I remember a conversation with Enrico Fermi after the war, a short time before the first hydrogen bomb was to be tested in the Pacific. We discussed this plan, and I suggested that one should perhaps abstain from such a test considering the biological and political consequences. Fermi replied: But it is such a beautiful experiment."

Figure 22 (p. 132) illustrates the principal relationship between basic scientific research, technology, and wealth. As I emphasized in the lecture on "Motivation in Science," science *costs* money, manpower, and material investments; it is technology (after appropriate innovation) that *makes* money and increases the wealth of the people (11). Most people in Europe or North America will appreciate (vaguely) the cultural values of scientific research; however, it is wealth and security, freedom from hunger, freedom from poverty, freedom from disease, and freedom from military threat that most people want from science. The presumed close connection between science and material welfare is the main reason why the public and why governments support science.

As BROOKS, a profound and sober expert in the field of applied science, writes (12): "I have little to say about the support of science for its own sake, for its intrinsic social and cultural value. Although there is no question that the public has demonstrated its willingness to provide such support, I doubt whether the intrinsic cultural value of science could be used to justify to the public or to politicians more than a small fraction of the present support for basic science in the United States, or indeed in any other major country of the world . . . This does not mean that the public is unwilling to support some very abstract science of no apparent usefulness, but the most persuasive justification for this is likely to be that science is a seamless web such that the 'useful' parts cannot prosper unless the apparently 'useless' parts are also well supported."

I want to consider some conspicuous features of the scheme (Fig. 22) and then return to the crucial question, science and people.

I have repeatedly emphasized that all genuine knowledge elaborated by basic research is potentially useful for achieving goals in accordance with the model for teleological action (see Fig. 11). It is the total stock of genuine knowledge that must be considered as the basis of technology, in particular it is that knowledge that is understood well and intellectually organized in empirical laws and in the paradigms of the different scientific disciplines. So-called technological discoveries or inventions can be understood as being based on available genuine knowledge (data, laws, paradigms), but as a rule they cannot be traced back to any *particular* discovery on the level of basic research.

In those cases where market research, social pressure, or an obvious need for a new product initiates the onset of "Research and Development" the level of basic research is often not influenced at all. Rather, the already available information from basic research (data, laws, paradigms) is being used as a reliable basis for the activity leading to successful innovation.

JEVONS points out that "only in very rare instances is it possible to pinpoint a specific curiosity-oriented discovery from which a wealth-producing application is derived" (11). Rather, it is the availability of objective data, laws, and paradigms and the existence of a body of experts well trained in research and in the spirit and paradigms of science that make innovation, development, and technological application possible.

The major drive comes from need. According to JEVONS "need pull" (along the solid arrows in our model) is much more important than the "discovery push." However, in those cases where large and rapid changes in the technology are involved (such as in nuclear power) the incidence of discovery push is greater. No linear model with a clearly defined starting point (discovery push model; need pull model) can describe the phenomenon that in fact leads to the advancement of technology and industry (13).

Feedbacks—indicated by dashed lines in Figure 22—complicate the model. For example, it is obvious that the engagement of capital on 'higher' levels (e.g., to meet the demands for various kinds of 'defensive investments' such as antipollution equipment) is directly competitive with research. This competition strongly affects the number of scientists involved in basic research, mission-oriented research, and innovation. It is difficult to estimate the amount of money actually spent at present by industry in the realm of research and development. For example, KORNBERG stated that "there is no industry based on technology today that spends less than 5 per cent of its product income on research and development" (14), while KNOX replied that "the petroleum refining industry, long considered to be a high-technology industry, spends considerably less than 5 per cent. Its R & D costs have usually amounted to one per cent or less of its product sales income" (15).

In any case a composed model, including multiple interconnections and feedback loops, is required to represent even the principal occurrences in connection with the development of science-based technology. Moreover, besides discovery push we are all familiar with *political* push and besides real (social) needs we know artificial needs as well. JEVONS concludes that "the overwhelming fact about innovation is the formidable complexity and variability of the circumstances surrounding it" (11). The complexity will be increased by the impact of the ethical questions, discussed in the foregoing section of this lecture, and by the very important fact that technological innovations take place in an atmosphere of strong competition between companies and nations. The creation of genuine knowledge including the paradigms of the age is a truly international enterprise (the *global* scientific community is no fiction!); technological innovations, on the other hand, are primarily *not* international. This cannot be changed. At least some of the highly industrialized countries depend desperately on their ability to sell their goods on competing world markets. For them effective science and technology has become an affair of State.

The British example makes clear that excellence in basic research and in research and development alone is not sufficient to guarantee national prosperity. Britain has been traditionally eminent in research and technological invention, but British industry collectively has never quite succeeded in meeting the rising competition from abroad in the period after the 2nd World War. The key problem is productivity. Productivity is the added value per man employed of the goods he works on. According to FLOWERS (16), in the USA a working population (1969) of 70 million produced annually an added value of about £350,000 million, or about £5000 per man on the average. In the UK, the corresponding added value per working man is only £1300. Since nobody can economically pay a man on the average a salary greater than his added value, the average British citizen was worse off than his American counterpart by a factor of about 4 in 1969.

I now want to consider three questions that transcend the model in Figure 22.

1. It is generally agreed that the need for foresight in technology assessment is overwhelming; yet foresight is always imperfect and needs to be constantly corrected by further experience. Moreover, to impose too heavy regulations from the very beginning on a new enterprise may have the undesired effect that the new technology will not be developed at all, although it might be urgently needed to replace an established technology

whose faults of operation and production have become obvious. This is the case with atomic energy plants.

2. In the past we have been confident that a technological solution can be found for nearly every problem, provided the data, laws, and paradigms of basic research are available in the particular field. It was thought that a body of trained experts would be able to overcome any difficulty or case of emergency that might arise, including energy production, food production, infectious diseases, and meteorological or geophysical catastrophes. But there are instances where this confidence is shaken: there is no technological solution in sight so far for the self-destructive population explosion in the Third World countries; the destruction of the commons proceeds at an increasing rate (pollution of fresh water, oceans, air; eradication of the tropical and subtropical forests; increase of the CO_2 level in the atmosphere; consumption or waste of nonrenewable resources); the threat of a global famine is very real despite the transient relief by the green revolution; the limits to growth in a finite world are obvious (17), even though computer simulation of the final collapse may not be precise in detail (18). The loss of confidence in the future is a major theme in discussions among leading scientists.

3. There are only a few countries in the world in which science really flourishes. There are too many countries—in fact the majority of the countries in the United Nations—where science is neither supported properly nor encouraged to grow. At present the developing countries possess some two-thirds of the world's population but conduct only 2 to 3% of its research and development—and the results of their efforts are of dubious quality (24). The tacit assumption can be made that the only motivation for science in the less developed countries is an immediate contribution to today's economy. As a matter of fact, these countries do not contribute significantly to the increase of genuine knowledge, nor do they contribute to the formulation of the paradigms of our age or to the large-scale solution of problems in the realm of "research and development" (see Fig. 22). As a consequence of this unfortunate situation, the rising investments (money, manpower, material goods) for the furtherance of genuine knowledge *and* for technological innovations are actually being made by a few countries only. As a rule, this investment *for the sake of all men*, which includes the support of graduate schools and research institutions, has not been appreciated by the Third World and by those countries that possess the major resources of raw materials (fossil fuels, minerals) required for present-day technology.

There is a real danger that the highly industrialized countries will be forced by unwise political price pressures *to sell* (rather than to give away) even the genuine knowledge elaborated with high costs in their research institutions and universities. So far some kind of control by political power over the spread of knowledge has been exerted only in the nuclear area and in some fields directly related to classified military projects. However, a similar control system could be extended to all basic science. Such measures would weaken and possibly destroy the global scientific community.

So far the truly international, global scientific community has refused to let hostilities between states break up the scientific community and interrupt its communications. However, the day could come when the conflicts of nations about vital resources have become so severe that the political powers will decide to forbid or at least delay the spread of genuine knowledge. Under these circumstances, knowledge obtained by basic research will become a political and economic weapon similar to crude oil, phosphate fertilizers, tin ore, or wheat. It would be a terrible weapon. In the words of the illustrious social scientist RAYMOND ARON (19): " . . . the sciences represent today the great human venture, the reason for working, the pride of modern civilization. Without them there would be neither cultures worthy of their name, nor nations capable of survival." Francis Bacon's axiom "knowledge is power" would be confirmed even with regard to genuine knowledge, if this unfortunate development cannot be inhibited, and basic science would be coupled to politics more strongly than ever before. I am afraid that this strong coupling is inevitable anyhow. Governments can no longer do without science, and science cannot do without the State, since the capital investments modern science requires can only be supplied by the State.

The scientist's dream of independence is a nostalgia. This does *not* imply, however, that freedom has become an illusion—freedom of thought and freedom of inquiry. We have shown previously that institutional and financial *dependence* and intellectual *independence*—freedom of thought and inquiry—are compatible as long as both partners, government and scientists, carefully observe the rules of their "contrat social." This has clearly been recognized by top administrators such as SALOMON (20) from the OECD: "In no part of the world can the political authorities dictate to science its procedures, the laws of its activity, its substance; it is not within the powers of governments to determine the form or the content of the scientific approach. The road to truth is as immune from political decision as truth itself; at a deeper level, truth . . . has its own authority which cannot be vitiated by the authority of the powers that be. The public authorities can possibly restrain its exercise by preventing its access to public discussion, by screening part of its results or by distorting their meaning, but there is no constraint or persuasion which can change what it has established, except the authority of the scientific discourse itself." However, SALOMON continues: "Even though the political authorities cannot tell scientists how they should search or . . . what they should find, everywhere they aspire to tell them what they should look for, inasmuch as they guide scientific manpower into this or that field of science rather than another." This is a fair description of the status quo. The close interconnection between science and State has become unavoidable; government and scientists have inevitably become partners.

I have ignored in this lecture the discontents with technology that fuel the antiscience movement of present-day countercultures. It is not my task to repulse the criticisms of the leading ideologists of the counterculture [such as Roszak, Goodman, Marcuse, Schumacher, or Hess and more recently HAZEL HENDERSON (21)] with regard to goals and organization of present technology. The relationship between science and technology (as described in the present lecture) remains the same, regardless of the kind of technology people choose in a modern, open, democratic, and necessarily pluralistic society. The eloquent leaders of the countercultures do very well as long as they criticize the abuses and giantisms of technology ["small is beautiful" (22)]; however, they fail completely in creating a credible and workable alternative. The chief deficiency in SCHUMACHER'S *Small is Beautiful* (22) is that it describes a number of utopian ideals without offering convincing signposts as to how they may be attained. The urban and rural idyllic utopias of the counterculture are hardly more than romantic visions. None of these constructions would survive in reality, if the framework (including tools, medicine, electricity, crop varieties, etc.) would not be supplied by "normal," highly efficient physical or biological technology. It is a terrible misunderstanding of reality to assume that any kind of "culture" would be compatible with largely undisturbed nature. Natural ecosystems in steady state produce no surplus, have no yield! Human culture, in particular agriculture, always implies a dramatic change of nature. The best that we can (and in fact should) achieve are man-made ecosystems with a relatively long half-life, which please our esthetic requirements. To hope for anything else is blank, irresponsible illusion.

Our ancestors did not consider the long-range side effects of technology on the environment and on society because they did not pay attention to them (they were—as a rule—concerned with their immediate survival) or did not have the knowledge to anticipate or prevent them when they invented medicine and agriculture or started to build the big cities. We realize now that the regressive phenomena can no longer be ignored and that a solution must be found, hic et nunc. We must pay very soon the tremendous bill that has accumulated during cultural evolution; otherwise the amounting interest will inevitably lead to a global default.

What can the present scientific community contribute? I agree with WEISSKOPF (23) who says "that to solve the natural science problems in our environment in its widest sense, we need scientists trained in pure science. . . . the attitude, the spirit, the state of mind which is created in the practice of basic research . . . is the basis of our ability to deal with nature . . . Our universities should not turn out environmentalists. Rather we must train geologists, physicists, chemists; we must train biologists in the basic sciences so that they can apply the skill, the attitude, the state of mind in which they were trained to those problems. Only then we will succeed."

Our prime task is to solve obvious problems. This can best be done by a highly trained person, trained in the techniques of a particular discipline and in the spirit of science. It is not our prime task as *scientists* to create or invent problems or to operate as a consciousness raiser by emphasizing particular *extrinsic* goals or values, *extrinsic* with regard to science.

13th Lecture

The Crisis of Science

At present, science as a cultural force is undoubtedly in a deep crisis. For several decades the scientist ranked high in the Western European or North American societies. This has changed (1). In our days the scientist and scholar seems to be an enemy of both nature and human society. Science, definitely, is on the defensive these days.

Science has been an inviting target for criticism over the centuries. Science began originally as a determination to rely on experience instead of on the ancients or upon ecclesiastical authority or pure logic. The countermovement was strong. I only want to remind you that in 1591 Galileo became so unpopular *for doing experiments* that he had to resign his position in Pisa and withdraw to Florence.

The present-day countermovement against science is pushed by academic humanists and journalists at a time when the genuine objective knowledge of science is required more than ever to develop a strategy for survival. To sacrifice science at the present point of cultural evolution would imply—nearly as a matter of course—the suicide of the human race. Why then has an attack on science been launched at all? Is it only because most intellectuals are illiterate in science and hence incapable of really understanding our science-based technology? But why did the public and some governments respond affirmatively to these attacks?

There is little question that the bulk of the laity and the majority of the college students in Western Europe and in North America are extraordinarily illiterate in science (2). As CASSIDY (3) has said, “many students seem not to believe in a reality independent of their wishes; they seem deeply ignorant of the relation between cause and effect; what they do not know seems not to worry them. They are ignorant of their own ignorance and seem not to think seriously of consequences. They are in a basic sense irresponsible.” I lay the blame for this not only on parents, outdated school curricula, and poor science teachers, but at least to the same extent on the intelligentsia who runs the mass media and in fact determines and manipulates public opinion. Most of these intellectuals are illiterate in science, totally and hopelessly disconnected from the main stream of scientific thinking. The grotesque fact is that they run a society whose existence is totally and irreversibly based on science and technology.

The present attack on science is largely misdirected. Science can be blamed neither for failures in goal-setting nor for the partial misuse of technology or for domination. The latter accusation with its strong emotional appeal is particularly unfair.

Even a brief glance at cultural evolution reveals that human ingenuity has invented a splendid repertory of elaborate legitimations and justifications for domination, both religious and secular, before or independently of the ascent of science. A fair and unbiased study of human history reveals further that science has improved not only material wealth and security, but also freedom: freedom from hunger, freedom from disease, freedom from domination.

Science as a cultural force, as an enterprise of the scientific community, has been the only means available to the human race so far in building a consensus of certified knowledge that goes beyond the subjective judgement of any individual (prejudice) or of any human grouping (ideology). Science can claim to have achieved a high level of objectivity, far higher than any other sector of human culture. This objectivity has become the basis for the wealth, security, and freedom we enjoy to a much greater degree than any generation in the past.

The misunderstanding of science by society and, in particular, by most of its intelligentsia, is a grave and depressing phenomenon. In the local newspaper (Northampton, Mass.) a headline reads "Science—once bulwark of objectivity—now meets with scepticism from public." In *Time* magazine, September 22, 1975, p. 59, you could read " . . . The human race may begin to fear its scientists to such an extent that it will take uncontrolled action against them. . . . When I think of a scientist, I think of intellectual curiosity triumphing over moral responsibility." These quotations stem from an article that comments upon the results of a readership poll, performed by the *New Scientist* and *New Society*, two British weeklies, in order to determine whether scientists see themselves differently from the way nonscientists see them. The results confirm the wide chasm between what C. P. SNOW called the "two cultures" (4). The *Time* correspondent concludes that " . . . As a matter of fact, the poll showed that a great many people who have strong opinions about 'the scientific community' today are not really familiar with it. Of the 20 scientists most frequently mentioned by name in responses to the survey, only 7 are living . . . the rest included such figures from the myth-laden past as Archimedes, Galileo, Marie Curie, Darwin and Einstein."

According to the *Massachusetts Daily Collegian* from November 12, 1975, the President of Columbia University, WILLIAM J. MCGILL, said that "during the next century, universities should concentrate on the study of mankind to help balance the destructive forces developed by science in the 20th century." Apparently, this was the tenor of McGILL's speech when he addressed the International Association of University Presidents and the American Association of State Colleges and Universities.

The ignorance and hostility of society's intellectuals in all matters regarding science is a particularly shameful phenomenon insofar as it is precisely this society which enjoys at least most of the material goods of science-based technology without hesitation. Except for a few intellectuals whose apocalyptic writings we may ignore, nobody seriously wants to place medicine back to the Middle Ages. However, society has become used to physical and biological technology to an extent that most people take it for granted. They have simply forgotten that these achievements do not exist as a matter of course, but must actively be maintained any time by a continuous strong effect of the scientific community. In 1954, the invention of a vaccine for polio was celebrated as the "end of a killer disease." Today hardly anybody outside the scientific community knows the names of Salk and Sabin, whose combined immunobiologic efforts have wiped out polio so thoroughly that there is now some fear among experts that laxness about prevention could bring the disease back (1).

As members of the scientific community, we must be deeply concerned about the antiscience movement and about the disturbed relationship between the scientific community and the public. As scientists, we need not only open communication and freedom from prejudice and political ideology; it is the economic *and* moral support from a flourishing society that makes inquiry possible. Science and society are closely interconnected in the modern world. A collapse of society will inevitably lead to a collapse of science. A breakdown of science will rapidly affect the material existence of a society that lives largely from the fruits of science-based technology. The situation we face at present is described by Holton (5): "New books that declare the supposed philosophical evil of science (above all, 'rationality') in florid counterculture language, are sure to be spread about by the book clubs and the paperback houses, and to give courage to those who look for popular reasons for decreasing the support for scientific research." It is not only the exclusive circle of the academic intellectuals of the counterculture, it is the members of legislature who read these books. THEODORE ROSZAK's books (6) can be found on many important desks and bookshelves.

How can an intelligent, knowledgeable, and devoted man like ROSZAK blame *science* for the failure to control man's selfishness and irrationality? How can he misunderstand the scientific subculture and particularly the scientific *rationality* so deeply (6)? What the prophets of the counterculture attack is a poor caricature of science. Take SCHUMACHER, whose *Small Is Beautiful* (7) has become a countercultural cult text. He says, "Science and engineering produce 'know-how'; but 'know-how' is nothing by itself; it is a means without an end, a mere potentiality, an unfinished sentence. 'Know-how' is no more a culture than a piano is music."

The counterculture's attack on science is [as HOLTON (5) puts its] "largely a fight against straw-men of their own making." But, unfortunately, the public and the decision-makers take the straw-men for real. Why? Again, it is at least in part the scientist's own fault! As Sir PETER MEDAWAR (8) observed: "Scientists are usually too proud or too shy to speak about creativity and creative imagination; they feel it to be incompatible with their conception of themselves as 'men of facts' and rigorous inductive judgements."

ANN PALM (9) appealed to the scientific community "to stop the notion that natural science is hard science to be done by hard heads on top of bodies without souls."

On the other hand, however, the essentials of "hard-nosed" science, careful definition, rigorous symbolic language, critical examination of statements, may not be diluted or sacrificed under the pressure of the counterculture. Rather, they are to be highly recommended in all *cognitive* areas. The *scientific* approach has been the only way so far to gain objective knowledge about the nature of things, and the *meaning* to our lives generated by the scientific approach could help us to overcome alienation and the ideologies of absurdity in which the young generation is mired.

In any case, the public image of science as determined by the mass media and by the paperbacks of the counterculture is desperately wrong. There, you find nothing about the nature of scientific discovery, about its motivation and innovative vigor, nothing about the structural beauty of the grand theories, nothing about the intellectual and emotional satisfaction the human mind may receive from the consistent rationality, harmony, and order of the universe and nothing about science as a cultural force, creating *meaning*, nothing about the happiness that comes from intellectual discipline and achievement.

The scientific community has responded poorly so far to the challenge, and the confusion and disenchantment of the wider public are still increasing. It seems that the understanding *within* the scientific community of the meaning of scientific discovery and of the nature of the scientific enterprise is too poorly developed to permit a forceful counterattack or at least an effective defense strategy in favor of the sciences. ROSZAK'S "challenge" is counterfeit; it has been the general ignorance of scientists in almost all matters outside their field of expertise that allowed this "challenge" to stand and to become influential. HOLTON, considering the position of the practicing scientist between philosophical Popperians (Karl Popper and his school) and the prophets of the antiscience movement such as Roszak or Reich warns: "One side condemns the scientists for being too rational; the other chides them for being too irrational. Caught in between, scientists, virtually without exception, pay no attention to either side, not even to defend themselves against grotesque distortions of what it is they really do" (17).

Snobism and a lack of courage and common sense contribute further to the constitutive weakness of the scientific community.

DETLEV BRONK (10) warned more than 20 years ago that

> Men of affairs and social influence need more knowledge of and appreciation of the traditions, ideals, and significance of science. Scientists are in part to blame for such lack of awareness . . . we have emphasized too much our discoveries and their useful applications. We have inadequately revealed science as a great intellectual adventure. Unless this quality of science is more generally comprehended, we shall be subject to adverse pressures that result from lack of understanding.

Moreover, scientists have contributed directly and frivolously to the development of the antiscientific attitude we face today.

1. In the golden sixties there were far too many promises and claims (1). Some of us even consciously nourished the "grand illusion" that every problem could be solved nearly instantaneously if we only wanted to solve it. The actual achievements in some fields (e.g., cancer research) are being compared now with the original expectations.

Some of us desperately warned that this avalanche of expectations should not be set in motion. The disappointment *had to* follow! The responsible scientist knew that the loss of modesty—i.e., the restraint in making promises and guarantees—would be accompanied by a terrible loss of credibility in the eyes of the public. Particular scientific successes, such as landing on the moon or the deciphering of the genetic code, have just seduced scientists into making exaggerated claims and promises such as "defeating cancer" or "solving urban difficulties." As BROOKS (11) rightly remarks:

> ". . . In each case . . . the relatively modest claims of the originators of a new intellectual technique tended to escalate in the hands of their disciples and followers."

A major problem has been that many scientists could not discriminate between problems that are accessible to a *technological* solution and those that require a change in attitude, propensity structure, or value system in order to be solved. It is the latter type of problem that creates the real obstacles, and it has been the misunderstanding of the structure and nature of *problems* that has led to the optimism that a global attack by the elite of pure scientists could *solve* the problems created by scarcity and overpopulation, racial tensions and economic pressures, crime and delinquency, interpersonal and intergroup conflicts leading to ideological intolerance and block formation, nationalistic chauvinism and hateful open war.

CASSIDY (3) said: "If now we could apply the effective management that brought the crippled Apollo XIII back to earth to the problem of our own 'Capsule Earth', the space program, justified as it is, would justify itself a thousand-fold more." Unfortunately, we cannot apply the skills, documented in the fulfillment of the space program, to an analogous "earth program," because the goal-structure of an earth program is ill-defined, whereas the goal-structure of the space program is clear and simple. Earth programs without drastic changes in value system and propensity structure usually lead to piecemeal solutions that are hardly more than relocated problems.

2. Some scientists have contributed to the *emotional* and *irrational* discussion about the negative side effects, the regressive phenomena of technology, without understanding the structure of teleological thought and action. They have openly misused their prestige as scientists in the political arena in support of a particular private or collective prejudice.

3. Publicly vented controversies among scientists, so often rancorous and shrill, have spoiled the image of the reliable, objective scientist in the eyes of the citizen. The Teller–Pauling confrontation has had this consequence in the past; the Jensen–Hirsch type of confrontation, which we can desperately watch at present (12), will possibly stir up the public still more, since "nature vs. nurture" with regard to human intelligence is a theme that concerns nearly everyone. As COTGROVE (13) puts it: ". . . the antics of individual scientists who parade their ideological preferences under the guise of scientific objectivity outside the controlling mechanisms of the community of science, . . . threaten the high standards of their calling."

In fact, public hearings on controversial matters at the intersection of science and society have usually been disastrous for the prestige of science. Of course, two scientists may disagree on a scientific matter that is really beyond the proficiency of science. It is the *behavior* of some scientists in the public forum that threatens the moral standards of the scientific community. The sad situation has been competently analyzed by WEINBERG (18): "When scientists express opinions on scientific matters in the public forum they are not subject to the sanctions that regulate opinions expressed in the usual channels of scientific communication. Because these traditional sanctions do not operate, the extrascientific debate often tends to be irresponsible scientifically: lower standards of proof are demanded in the public than in the professional debate, and half-truths are too often perpetrated on the public by scientists. . . . "If scientists allow themselves the right to speak sloppily on science in the public forum, I think that this habit could gradually encroach upon the scientific forum." I want to add that irresponsibility of this kind and caliber will rapidly deteriorate the trustworthiness and the prestige of science in the public. We will destroy the basis on which we depend.

4. Some leading scientists have usurped the position of decision-makers and have promptly failed. Obviously most people in the Western societies do not like the scientist playing the role of a final arbiter with regard to legislature, in particular if unpopular decisions are being introduced. The naivity of many scientists, suddenly involved in public affairs, is sometimes overwhelming. Some of them even went so far as to assume that the public would respond fairly and consistently to reasonable and necessary measures. Even a minimum knowledge about human history and about human nature tells you that this assumption is idle.

I hope that the public disillusionment with the performance of science, however unfair it is, will have three strong effects on the scientists:

Firstly, this experience will show the scientist where his limitations are. Science is not a kind of religion, and it is not metaphysics. A scientist is neither a priest nor is he a philosopher in Plato's sense. He cannot easily satisfy the human need for metaphysical certitude. It is not easy to find "gnosis"[8] [to use ROSZAK'S term (6)] and salvation in the statements of science.

Secondly, the scientist will realize again that he is primarily a member of the *scientific* community and that—as a rule—he is not prepared for excursions into the political arena. In particular, the experienced scientists must learn to abstain from alliances between individual scientists and politicians to push pet programs. PEKERIS, after a depressing and pessimistic analysis of failures of individual physicists sitting and acting on political panels, formulated the lessons to be learned from these events as follows (14):

1. That scientists sitting on government panels must be ready to devote more time to the task, or must refuse to participate.
2. That the community of physicists must assume collective responsibility for the operation of these panels and should certainly seek to be informed about them.
3. That this clearly requires expenditure of more time on public problems by each physicist than he has allocated until now.

Thirdly, the scientist will learn the lesson that society, in particular its academic humanists and journalists, is an unreliable partner from whom you may not expect the minimum amount of fairness usually considered as indispensable among the members of the scientific community. A society that forgets almost instantaneously about the positive aspects of technology, about electricity and food, antibiotics, fertilizers, polio vaccine, insulin, contraceptives, solid-state technology, is a strange partner. As scientists we must realize that we deal with a society which is so snobbish that it is

[8]"Gnosis," according to ROSZAK (6), is real knowledge, which seeks "that meaningfulness of things which science has been unable to find as an 'objective' feature of nature."

regarded as a sign of bad taste if you remind somebody of the life style and suffering, of the pain and back-breaking labor of our grandparents. It is hard to think of any area of modern welfare that cannot be traced in some way to science, but any statement in this direction is regarded as obscene by our leading intellectuals.

Does the hostile part of our ungrateful populace really want to get rid of the scientific community? Do they really believe that genuine knowledge would be dispensable? This question may be answered by an underground joke I heard in Russia. A person asks a question to radio station Eriwan (this is a fictive radio station in the capital of Georgia to which one may address all questions that bother a Soviet citizen): "Shall I become a communist." The answer is: "In principle, yes; but we advise you to become a communist in a capitalistic country." The antiscientific consciousness-raiser in our society wants to get rid of science without sacrificing the comfortable privileges of our science-based culture. I have never met a sincere critic of the scientific community who was prepared to die, tomorrow, because of appendicitis, diabetes, pernicious anemia, or malnutrition. However, I have not met any critic of science either, who had ever seriously thought about the prospect that the breakdown of the structure and morale of the scientific community would rapidly lead to an impetuous breakdown of our culture and to a gigantic and fatal disaster for most of us.

The eruptions of the occult, mystical, subjective, the well-publicized offensive of the antiscientific irrationalism by some of the leading lights of the Western European and American "intellectual community," happened (and could possibly only happen) within the framework of an economically successful but undisciplined society. It was an emotional response to affluence, waste, and permissiveness. (1). We recall that the unique success of the scientific community (with regard to obtaining genuine objective knowledge) requires intellectual discipline, some degree of renunciation, constant struggle for excellence, and some degree of conformity. The new prophets promised the easy way to permanent happiness and satisfaction, by passing the hard intellectual pursuits. They promised knowledge without strong intellectual hardship. This phantom will probably disappear when affluence disappears and the real problems of our present time become demanding. However, it is questionable to what extent and at what rate science as a cultural force will recover.

A particularly serious point in this context has been the identification in the public of the *natural* sciences with the so-called *political* and *social* sciences. Any responsible social scientist will admit that genuine objective knowledge that could function as a reliable basis for practical measures is very scarce in his discipline. For a while (in the sixties) the illusion could be maintained that the social sciences did in fact *have* the knowledge to suggest solutions for any demanding problem of society. In fact, the social sciences have greatly overestimated their potential ["we were giving, and demanding the right to give, advice as giants when we were still pygmies" (1)]. The failure to meet the actual problems such as recession, inflation, urban social problems, excessive misuse of social welfare, etc. was unavoidable. As a result the trustworthiness of the social and political sciences declined rapidly and with it the trustworthiness of the whole scientific community, including the natural sciences. It is an old experience that a counterreaction does not discriminate between those who are more and those who are less guilty.

The switch of social "scientists" to Marxist ideology on a large scale in Western Europe and sometimes even in the USA finished by and large the prestige of this discipline as a *scientific* field, subject to the normative rules of the scientific community. But the harm done to all of science in the public will persist.

Our immediate task is to restore the confidence of the public in the trustworthiness of the hard core of the sciences. If this can be accomplished, science will attract once again the best-endowed minds for the pursuit of genuine knowledge, and it will reattract the moral and financial support of the public. The *primary* goal of science must be restored in the minds of scientists and laymen alike: the search for knowledge for knowledge's sake. It is not *primarily* the purpose of science to "advise governments, save mankind, make public policy or build empires" (1). Recently an experienced, sober journalist asked the members of a scientific community, who were debating the impact of science on society; "do you believe that good scientific research can be accomplished with high-minded sentiments" (15). However, *it is* part of the responsibility of the scientist to be ready, at any time, to contribute to the rationality of political decisions by thinking in terms of if/then statements and by advising the politician in accordance with the cooperative model that I have outlined in the Prologue and reconsidered briefly in the preceding lecture.

Another grievous symptom of the crisis of science has been the inability of the scientific community to anticipate and to deal with the rapid changes of the growth rate. The growth of science in the leading countries has been exponential over a considerable period of time. This has changed now. In the near future, we will be faced with zero growth or even with negative growth rates, at least in the realm of basic research. This state of affairs had to be expected. No component of society can long continue to grow at a rate greater than the rate of growth of the society as a whole. But, nevertheless, we have not seriously analyzed the equilibrium situation ahead of time, and the inevitable consequences are not fully clear to us. In fact, the system 'scientific enterprise in its relationship to society' is only partly rationalized, and we can only guess about the psychological consequences within the scientific communities of a rather abrupt swing to equilibrium conditions. The magnitude of the impact to be expected from a transition to equilibrium has so far been seriously underestimated by most scientists who grew up under conditions of nonequilibrium growth. One aspect that can, with relative ease, be treated quantitatively concerns the number of science graduates required per year under equilibrium conditions. MARTINO (16) calculated in 1968 that the scientific staffs of U.S. universities were already larger than the staffs that would be required in 1975 if science came into equilibrium with society in 1968. ("Equilibrium" in the present context means that the growth rate of science is not greater than the growth rate of society as measured, e.g., in terms of gross national product.) Equilibrium conditions—the best that we might expect in the foreseeable future—imply that U.S. universities will find that they are required to turn out a much smaller amount of graduates, most of whom will replace losses rather than fill new posts opened by growth.

MARTINO (16) makes the point: "that the non-equilibrium growth of science over future generations has become an unquestioned assumption underlying many of the practices, customs, and habits of science as an institution, and of individual scientists. Many of these practices, customs, and habits may have to be modified considerably in the transition to equilibrium. . . . Because of the deep-rooted nature of the assumption of nonequilibrium growth, the nature of some of these modifications may not even be apparent until the situation has reached the crisis stage." We *have* reached the crisis stage, and we are indeed largely unprepared. We simply do not know how to handle university departments or research institutions efficiently under conditions of equilibrium or dwindling funds. We will learn, but it will be a painful experience. In the end, hopefully, it will be a sound experience.

A major *psychological* problem will arise from the fact that a reduction in the rate of expansion of science implies a reduction in opportunities for advancement of younger scientists. The average age at which scientists reach top positions such as department heads, chiefs of research groups, adviser to the government etc., will increase. To the extent that creativity is associated with youth, the result will be a decline in average creativity.

Considerate observers of the scene warned years ago "that science has reached the point where science policy makers must start considering the transition to equilibrium and, in particular, must start thinking about how to cushion the shocks which will accompany this transition" (16). However, little has been done by the scientific establishment to make the transition to equilibrium as smooth as possible: a nearly inexcusable neglect.

Still worse, a solid defense strategy in favor of the *budget* of science has not been prepared. Lamenting ad hoc articles in journals such as *Science* and *Nature*, although well-aimed, do not suffice. We all knew that science could not expect to be exempted from all consequences of the governments' fiscal crises. Instead of explaining, justifying, and thus defending the matter of science in a comprehensive "philosophical" manner, understandable for the politician and convincing for the educated citizen, we have left the field of public relations—in a combination of narrow-mindedness and arrogance—to our ideological and political critics who have worked tirelessly over the years to undermine and ridicule the scientific enterprise. Our critics have been quite successful. Many politicians and the bulk of laity really believe that at least curiosity-oriented basic research is a kind of superfluous extravagancy. This is the most serious factor in the present crisis of science.

14th Lecture

Science and Values

Values are indispensable elements in every teleological action (see Fig. 11). We all have values. A value can be considered to be the rationalization of an interest, a need, natural or cultural, an expectation or a subconscious drive. Our value system is thus tied up with the very essence of human nature. Values guide our decisions in goal-setting, the selection of means to reach the goal, and the assessment of risks and undesired side effects. It is our value system that determines the way we plan, act, perform, and regret. In particular, it is decision-making of any kind that is inevitably connected to a value system.

We recall from a previous lecture (see p. 42) that a statement is true if, and only if, what it asserts to be the case actually *is* the case. When is a decision right? There is no general answer since there is no general agreement about right and wrong. At first sight, the answer of the ethicist sounds convincing: "A decision is right if, and only if, it respects the intrinsic worth of all human beings" (1). However, what is "the intrinsic worth of all human beings" explicitly, in terms of our behavior, hic et nunc? Even the formulation: "Do not violate human dignity when making decisions" can hardly be operationalized, since it requires some kind of agreement among human beings that does not always exist. One thing we know, however, is that there can be no right decision without *knowledge*. Genuine knowledge is an indispensable prerequisite for right decisions and right actions. On the other hand we have emphasized in the Lecture on the Ethics of Science that knowledge is thought to be *good* in an ethical sense. Hence, truth and goodness are interrelated.

Figure 11 implies that values have a causal impact. This implication can be challenged. One can hold the view that values are in fact an epiphenomenon, a by-product of action. From this point of view, we infer our values and our propensity structure post factum from our behavior. We create values (ethical entities) to justify *in retrospect* our actual anticipation or action. As WILSON expressed it in an effort to answer the question: what is the main thesis of sociobiology: "It is that all forms of behavior in animals and man—including altruism, religious activity, ethical decisions—can ultimately be explained by genetic evolution, which is based on natural selection" (2).

Even though we might agree with Wilson, it is "practical reason" (in Kant's sense) that advises us to consider values as *determinants* of teleological action. The problem of whether or not values can in fact be considered as determinants of actions is similar to the problem of free will, which we have already discussed on two occasions (pp. x and 86). In the following, I prefer to consider free will and the intervention of values in causal chains as an *indispensable* fiction in human life and as an indispensable aspect of human distinctiveness and dignity.

ROKEACH (3) defines a value conceptually as "an enduring belief that a specific mode of conduct (instrumental value) or end-state of existence (terminal value) is personally or socially preferable" to its opposite or converse. A value system is an enduring organization of values along a continuum of relative importance (4). It is difficult to deal with values. As BRONOWSKI has noticed: "The concepts of values are profound and difficult exactly because they do two things at once: They join men into societies, and yet they preserve for them a freedom which makes them single men" (5). It has been the traditional task of "practical philosophy" to deal with the justification of values and (legal) norms, in particular values and norms for action. In LORENZEN'S words: " . . . moral philosophy has the task of formulating principles which allow us to work on merely subjectively 'given' wantings and so to discipline them that our decisions about how to act are justifiable" (6).

ROKEACH (3) preferred an empirical approach. He had respondents rank an alphabetically ordered list of 18 terminal values (e.g., a world of peace, pleasure) and then 18 instrumental values (e.g., ambitious, logical) "in order of importance . . . as guiding principles in your life." In his book, ROKEACH reports data obtained mainly from a national sample of American adults in 1968. Respondents completed the value survey quickly (in 10 to 20 min), and considered it thought-provoking and no invasion of privacy (4). It will suffice to quote just a few of the results obtained from this survey in order to illustrate some *group* differences within our society in ranking values (4). Males and females agreed in ranking *a world of peace*, *family security*, and *freedom* as most important, but males tended to rank a *comfortable life* substantially higher and *salvation* lower than females. The poor and uneducated ranked *clean* and a *comfortable life* higher, and *logical* and *a sense of accomplishment* lower than the affluent and highly educated. Whites differed from blacks most clearly in caring less for *equality*. Value differences appeared throughout the life cycle, from 11 years to over 70. *Wisdom*, e.g., was ranked highest in the college years, being lower in relative importance for both older and younger groups.

Regardless of field, academicians valued social and intellectual stimulation and achievement more highly than other educated Americans, but ranked religious and self-restrictive values lower.

The basic questions of ethics remain unsolved by this kind of an empirical approach. What does the ranking mean to the individual? Is there a more basic value system that is used by the respondents in ranking the values offered by ROKEACH in his value survey? Is there any other source of ethics besides survival needs or is the amazing variation of the value system within a "free" population comparable to the tremendous diversity of human beings with regard to morphological, physiological, biochemical, and mental traits? Let us briefly consider this question. Since the wide genetic variation within existing human populations seems to be fully consistent with the principles of Darwinian evolution (7), there is no need to postulate a source of spiritual values. Rather, the presupposition suffices that our value system exists because it has had survival value during genetic evolution. However, the origin of some "values" (such as the egalitarian ideology), which can hardly be reconciled with the nature of man as a result of genetic evolution, deserves special attention. I presume that these "values" must be considered to be cultural exaggerations of an underlying evolutionary principle, in the present case "inclusive fitness" (see p. 199).

Variation in man has the consequence that a given list of values will always be disputed as long as there is no externally enforced uniformism, no *ruling* ideology, no dogma that *must* be adopted by the people. However, there are obvious limits for diversity even in an open free society. A given value system must be accepted by the members of any political grouping at least to an extent that permits the living together in the same commonwealth (minimum common value system). Since at present we depend on cooperation on a *global* scale, a minimum common value system must be defined, accepted, and practiced by all people on this planet. The Charta of the United Nations could serve as a basis for a minimum common value system, but it has, in practice, never been accepted by all nations. The fact that physicist Andrei Sakharov's application to visit Oslo to receive his Nobel Prize for Peace was refused in November, 1975 by the Soviet government, documents once again that freedom of thought and speech, freedom of press, freedom to criticize, and freedom to travel cannot be conceded as a fundamental value by the totalitarian marxist countries.

In the field of elementary welfare most political groupings in our society accept a canon of basic values as indisputable. This minimum common value system includes health, education, adequate housing, equal opportunities, equal rights. However, this does not necessarily mean that there is—as a matter of course— an unquestionable common good for all people in an open, pluralistic society. In fact, common values such as the above-mentioned basic values must be conceived as social compromises that have become meta-stable, at the most. Thus, common values are vulnerable, sensitive entities that deserve careful and considerable treatment. We can only live in freedom as long as the framework of values that permit an open, free society is not seriously shaken. Ridiculing or disregarding traditional values and tearing down moral institutions without offering a workable alternative is dysfunctional in getting constructive change accomplished. Certainly, the value system of an open society must remain flexible, and there should only be a minimum of restrictions for the individual. However, we must realize that no person can have complete freedom, since all persons need some freedom.

The minimum common value system of a society must be defended against moral anarchy. To sacrifice a common denominator of values in favor of blind anarchy would destroy human culture. At least temporarily the minimum common value system of a society must be considered as a set of largely indisputable paradigms. Otherwise, a cultivated humane society, necessarily based on cultural *tradition* and moral *orientation*, could not be maintained. Freedom as a prime value does not imply that we can afford to doubt everything all the time.

We recall at this point that the truly international, *intrinsic* value system of science (the ethics of science) is compulsory and largely independent of cultural, national, or political boundary conditions. In fact, scientists of different cultures, e.g., scientists belonging to a marxist society on the one hand and to a liberal, pluralistic society on the other hand, will easily get together and, as long as they stick to their science, have no great difficulties in discussing the advances in their particular field. If there is a difficulty it is because the science has been itself polluted by *political* ideology as in Soviet physics during the antirelativistic and anti-Kopenhagen campaign or in Soviet biology during the Stalin–Lyssenko era. The fact that scientists agree on the ethics of science does not imply, of course, that they agree on *extrinsic* values, extrinsic with regard to science. Actually, scientists from different cultures and even scientists within the Western societies may substantially disagree about social provisions, about what society ought to be up to, the role of individual freedom, and other evaluative questions of this kind. Obviously, science cannot tell us convincingly what ought to be.

Moreover the rise of science has contributed to the crisis of orientation and to the conspicuous loss of security. As DAVIES (8) pointed out rightly, " . . . science has taken over, with great success, many areas where our previous beliefs invoked the supernatural to explain natural phenomena. It has, therefore, shaken our belief in the supernatural as a source of guidance. It seems to me that perhaps much of our current crisis arises from this loss of a set of stars to guide oneself by. The world hasn't found a substitute for the supernatural as a basis for a generally accepted code of behavior." The unspoiled human mind will respond to commandments because we need a set of coherent values for orientation. What the young generation in fact demands and needs is a new kind of religion, a new kind of security. Sometimes, even the wildest ideologies will readily be accepted on our campuses because they offer a holdfast and a compass. I think that we must return to persistent moral institutions that offer orientation. We must make strong demands of consistency and coherence in moral thought, in order to make the value system attractive for the bright and critical human mind. In brief, the value system must satisfy the emotional needs *and* the critical, analytical reason of our youth. Otherwise, the common basis of values will become too small, the generation gap will widen further, and finally become self-destructive for the Western societies.

I admit that perfect consistency and coherence in moral thought cannot be achieved. In practice, any value system will contain incompatible values. This can hardly be avoided since value systems in real life cannot be expected to be fully consistent with formal logic. The notion of tragedy in the work of our great poets refers to this irresoluble conflict of right and right in the history of the human race (9). A somewhat trivial problem may illustrate the inescapable incompatibility of values in our daily life: freedom and safety in traffic. We obey traffic lights (on the whole) despite the fact that basic freedoms are impugned. We consider the possibility of living together in the streets and on the highways as a *higher* value than complete freedom. As I already pointed out, no person can have complete freedom, if we want all people to have some.

Is there any function of science in the value system of human societies?

In philosophical ethics, it has been called a fallacy (since Hume at least) to suppose that a conclusion about what ought to be can be deduced from premises about what is, was, or will be. In SIMON'S (10) words: "The theory of evolution does not tell us what is good or bad, in any ethical sense; the evolutionary process itself is non-ethical."

I agree that it is not possible to derive a compulsory system of ethics from evolutionary theory, comparative ethology, or from any other branch of biology. However, I think it is possible (and necessary, as I firmly believe) to derive antivalues from biological knowledge. By antivalues I mean statements about things that may no longer be done or believed, if human welfare and the long-term survival of the human species are accepted as universal goals. Given these promises (which are debatable, of course), we can indeed not say what is "right," but we can state what is "wrong." The following two case studies are supposed to exemplify the critical function of science in the process of reshaping our minimum common value system as part of a "strategy for survival."

However, for the sake of fairness I want to emphasize at the end of this introduction that the impact of science on society seems to have improved the actual human behavior. It is an historical fact that the ascent of science has been *correlated* with a conspicuous progress in *humane* behavior. In the words of the philosopher DUNHAM (1): ". . . a certain large area of humane behavior has already been won. Time was (in Renaissance Florence, say) when you had to go armed when you left your house, the house itself having been built as a fortress—battlements, watchtower, and all. You can now live much of your life (some people live all of theirs) without being irresistibly tempted to harm your fellow men. A great many people have never killed or maimed anybody and they were not persons of weak passions either. We have therefore a broad social ground from which to advance toward the civilizing of all life. What we have yet to learn is that *groups* of men must behave with the same decency that individual persons now show one another. . . ."

The Ideology of Egalitarianism. Egalitarianism is based on the doctrine that "all men are created equal." Is this a normative or a factual statement? Is this assertion consistent with our *knowledge* about man? What are the relevant biological facts? Within a single, sexually reproducing species individuals differ in almost every trait (characteristic). This phenomenon is called variation. Variation has always fascinated the careful and thoughtful observer of nature and society. In 1636 Sir THOMAS BROWNE wrote (7): "It is the common wonder of all men, how among so many millions of faces, there should be none alike." It is not only the morphological traits that show a high degree of variation in man: rather, variation is also characteristic for physiological traits (e.g., blood pressure) or psychological traits (e.g., the operationally defined I.Q.).

The question has been to what extent the total variation of a particular trait in a biological population is due to genetic variation and to what extent variation in man and other species is caused by environmental factors. Means have been devised to determine "heritability." This is a measure that is defined as the proportion of the total variation of a particular trait in a particular population that can be attributed to the action of genes acting independently of each other. Since some genes may interact with one another and with the environment in complex, nonadditive ways, heritability measurement tend to underestimate the genetic component of diversity (7). Despite this tendency toward underestimation, the calculation of heritabilities for some human traits yields high values. Human height, e.g., seems to have a heritability of from 50 to 80%. It should be emphasized that these values apply only within the Western societies in which they were measured. The heritability of a trait may vary from one population to another, and it certainly depends on the degree of variation in the environment (7). A very important point is that a monotonous environment tends to increase heritability. If the environment is monotonous *and* favorable for the development of a particular trait nearly all variation will be genetic variation. In other words: heritability tends to be very high under these conditions. Therefore, increasing equality of opportunities in a society with a well-functioning educational system will strongly increase heritability of mental and intellectual traits.

What is in fact our present knowledge about heritability of essential traits in human populations? (I refer mainly to Western European and North American populations since these have been mainly investigated.) Modern methods of investigation, in particular protein separation by means of gel electrophoresis to study protein polymorphism, have led to the conclusion that people differ genetically far more than was once thought (7). "Two people chosen at random from the American or the British population will differ at hundreds of chromosomal loci, and possibly at thousands. These differences are not trivial; they affect, or in the past have affected, our survival and reproduction. Many of them are associated with differing susceptibilities to disease, and if those associations can be analyzed, medicine will be able to take account of them" (7). To deny this knowledge for the sake of the egalitarian ideology would be very unwise for the sake of man. The maintenance of an ideology that is not consistent with the facts of human nature would be deeply inhuman. I will briefly consider two aspects:

1. Human polymorphism has become the subject of the new medical science of pharmacogenetics which studies differences between individual reactions to therapeutic drugs. From the point of view of the Hippocratic Oath, it must be considered to be a crime to ignore polymorphism with regard to drug responses once the basic knowledge is available.

2. It seems unlikely that the observed morphological, physiological, and biochemical diversity could fail to have an effect on our mental performance and on our behavior. Indeed, within-population heritability estimates for I.Q. usually range from .60 to .85. For present-day populations of European origin the most probable heritability estimate with regard to I.Q. seems to be close to .80 (11). As CLARKE (7) puts it: "The existence of this diversity might encourage the search not for the ideal social or political system, but for the ideal array of social and political systems. We should, perhaps, ask for polymorphism in our institutions to match the polymorphism in ourselves." In any case, we have to face the fact that human beings differ greatly in their genetic endowment. It would be irresponsible to devise or maintain educational institutions based on the fiction that this genetic inequality among human beings does not exist.

Unfortunately, the notion of genetic diversity has encountered a good deal of opposition among egalitarians who fear "that its recognition will discourage efforts to eliminate *social* causes of educational failure, misery, and crime. Accordingly, they equate any attention to genetic factors in human behavior with the primitive biological determinism of early eugenicists and race supremacists. But they are setting up a false dichotomy, and their exclusive attention to environmental factors leads them to an equally false social determinism" (12).

This opposition of the egalitarians to scientific knowledge could be taken as the usual resistance of the defenders of beloved ideologies against the progress of scientific knowledge. We recall the opposition of theologians a century ago to the theory of evolution, which they considered a threat to the foundations of public morality. It has been surprising, however, that even some scientists who should know better, openly *discourage* the public to accept the notion of genetic diversity. As a sad example, in a recent (1975) "NOVA" program on Public Television, a distinguished population geneticist "denied the legitimacy of human behavioral genetics, scorned the belief that musical talent is inherited, and even minimized the contributions of genetics to agricultural productivity" (12). While the *social* motivation of scientists in favor of egalitarianism might be honorable, their behavior in public cannot be conceded by the scientific community, since they misuse their reputation in some fields of expertise to advance publicly their private opinions on behalf of science, although their opinions are *not consistent* with our present scientific knowledge. This may have tragic consequences for genetics as a science as well as for the welfare of people (I always presume that neither religious nor political fervor can command the laws of nature and that human life must be in accordance with the nature of man, if it is supposed to be a life in dignity and freedom).

> As Davis (12) pointed out: " . . . one can easily visualize an American Lysenkoism, prescribing an environmentalist dogma and proscribing or discouraging research on behavioral genetics. But such a development would deprive us of knowledge that could help us in many ways: for example, to improve education (by building on the diversity of individual potentials and learning patterns), to decrease conflicts, to prevent and treat mental illnesses, and to eliminate guilt based on exaggerated conceptions of the scope of parental responsibility and influence.
>
> In the continuing struggle to replace traditional myths by evolutionary knowledge the conflict over human diversity may prove even more intense and prolonged than the earlier conflict over special creation: the critics are no less righteous, the issues are even closer to politics, and guilt over massive social inequities hinders objective discussion. What the scientific community should do is not clear. At the least we might try to help the public to realize the value of scientific objectivity, separated from political convictions, in understanding human diversity.

I feel that we should try, in addition, to analyze as clearly as possible the situation into which a few inconsiderate egalitarian do-gooders have maneuvered the scientific community. I think that the basic misconceptions are the following:

1. The statement that people within a population differ in nearly every trait, including behavioral traits, does not imply any statement about values. These statements are scientific statements (if they are true) and are supposed to describe a state of affairs in the real world. They are neither normative nor moral statements.

2. Human dignity is correlated neither with the particular shape of my nose, with the color of my hair or my skin, nor with my height, my musical talent, or my particular IQ score.

3. The statement that two populations differ more or less in some gene frequencies is a scientific statement (if it is true). The use of this statement by racists has nothing to do with the statement itself. Every scientific statement can be misused. You cannot fight racism or any other kind of private or collective prejudice by simply forbidding science. The authors of a recent, meticulously written, most comprehensive critical and balanced review of the race–IQ issue have reached a similar conclusion (11): "We consider it quite likely that *some* genes affecting *some* aspects of intellectual performance differ appreciably in frequency between U.S. racial-ethnic groups—leaving open the issue of what groups, which aspects, and which direction of difference. Thus, we consider it most unwise to base public policy on the assumption that no such genetic differences exist. If someone defends racial discrimination on the grounds of genetic differences between races, it is far more prudent to attack the logic of his argument than to accept the argument and deny any differences. The latter stance can leave one in an extremely awkward position if such a difference is subsequently shown to exist."

I have repeatedly suggested that the high genetic variability among people should be considered a predominantly positive phenomenon (13). Admittedly, genetic variation among people has been a cause for envy and hate all through history; however, it has also been a decisive factor for the fascinating diversity of cultural expression and for the multiplicity of human efforts and achievements in art, literature, philosophy, science, and technology. Moreover, genetic diversity has been the basis for the uniqueness and dignity of every human life. To sacrifice the wonder of human individuality for the sake of egalitarianism would imply sacrificing the essence of human culture for the sake of an awkward ideology. I believe that the maintenance of human and cultural diversity is one of the highest values. Without diversity, our life would be colorless, poor, and boring.

Any *humane* society must guarantee equal opportunity and justice for everyone. This implies that all members of a society should have equal economic and political opportunities, and equal access to education and other general social benefits and services. But no society can assume or guarantee an equal distribution among people of the ability to make use of the equal opportunities. The social and human postulate of equality of opportunity and before the law is *not* based on a biological correlate. Rather, every human being, except monozygotic twins, is genetically unique, and this genetic uniqueness applies to behavioral and mental traits as well as to morphological, physiological, or biochemical traits.

It has been suggested (I was told) by some radical egalitarians that genetic differences be compensated for (to some extent) by *unequal* treatment. One could give less education to the bright and more to the less bright. While this barbarous procedure could perhaps counterbalance to some extent the differences in genetic endowment in intellectual traits, an equal phenotype could not easily be obtained with regard to physiological or morphological traits. What could be done in an egalitarian society against the natural beauty of a woman or against the healthy and vigorous appearance of a man? Would the introduction of a beauty tax take care of this problem? This admittedly ridiculous example is supposed to illustrate the major point: there are obvious limits to social envy (14). It cannot be the goal of a cultivated society to eliminate differences between the members of the society. This objective could only be achieved by suppressing the development of those people whose genetic endowment is beyond the average. Leaving aside the moral problem, no modern society could afford to waste its talents this way. Intellectual capacity is the most valuable resource in our endangered world. To waste it for the sake of an awkward egalitarian ideology would be more than irrespondible. As NORMAN HACKERMAN, President of Rice University, has noticed in a recent editorial in *Science*, " . . . talented individuals are not too plentiful and . . . talent should be nourished where it appears. In other words, it is unwise to disregard the real differences in intellectual capabilities, both for the individual and for society. . . . So recognize elitism in education for what it is: the opportunity for creative individuals to pursue intellectual goals and ideals somewhat beyond the boundaries confining many of us. If we destroy this environment of creative endeavor, whether it be scientific, artistic, or literary, we will have lost a great gift for humanity" (15). The only thing a humane society can *and must* do is to guarantee equal respect, equal dignity, equal rights, and equal opportunities to every citizen, irrespective of his or her genetic endowment. This implies that equality of opportunity in employment should be assessed in terms of an employee's ability or trainability to do a job and not his racial-ethnic membership.

Growth as a Supreme Value. Population growth and concomitant economic growth have been regarded as supreme values for a long time. Why? To strive for growth is part of our propensity structure. We subconsciously dislike belonging to a dwindling population or watching our economy decreasing its gross national product. This is biological heritage! Under conditions of natural evolution a strong propensity to grow, to expand, to conquer new territories and niches must have been essential for the survival of a population. Therefore, "growth" as a value became part of our genetic heritage—a "value" that has now become not only awkward and obsolete, but also dangerous and self-destructive. Furthermore, the Judaic-Christian *cultural* heritage as well as the marxist doctrine has favored the notion of growth. Thus, we are confronted with the power of a strong genetic as well as cultural tradition if we suggest a swing toward a steady state or equilibrium society (16).

Is growth—from the point of view of science—compatible with the notion of further survival of the human species? The notion of survival implies that we develop and adhere to a life-style that is compatible with the long-term function of the ecosystems (natural and man-made) and with the resource endowment of the earth. The basic issues are simple. Since the earth is a closed system (except for radiation), growth cannot continue indefinitely. If growth continues, the collapse of the system is only a question of time (17). Most experts agree that the world is already heavily overpopulated. Further population growth will rapidly outdistance food supply, leading to wide-spread starvation and continued environmental degradation. The present situation (December, 1975) shows all signs of a crisis. Of the 115 countries for which reliable trade data are available, all but a few now *import* grain (18). The food comes from North America (approximately 92%) and Australia-New Zealand (approximately 8%). The experts warn that the evergrowing dependence on North America by the rest of the world cannot continue much longer. A single failure of the grain harvest in North America would inevitably lead to world-wide starvation. Undoubtedly, population pressure for food, energy, water, and land resources is significant at present with a population of 4 billion human beings. With the human population projected to increase to 7 billion within 25 years food shortages and energy, water, and land limitations will become critical. Protein supply is particularly in danger. Even if we move toward greater consumption of vegetable protein, protein production would have to increase *substantially* over the next generation. In an excellent article on "Energy and Land Constraints in Food Protein Production," PIMENTAL et al. (19) point out that "these increases are questionable because: (i) shortages of land, water, and energy resources already exist and these shortages will intensify as the human population continues to increase; (ii) further ecological degradation of land, water, and vital biological resources will tend to reduce the productivity of the agroecosystems; and (iii) crops have physiological limits in their ability to respond to increased amounts of fertilizers and other crop production inputs." The authors conclude their scholarly article with the lapidary statement: "Clearly if man does not control his numbers, nature will."

Population growth is indeed the fundamental problem. If it cannot be solved, we will not have a chance to solve the others. The population explosion not only destroys the biosphere, it also destroys, irreversibly, the genetic potential of evolution, the wealth of animal and plant species, and it destroys human culture. There can hardly be any doubt that population explosion—if it continues without significant check for one or two generations—will terminate and finish off cultural evolution (20).

Why was this gloomy prospect not recognized earlier? Why was

Malthus not taken seriously? Apart from ideological commitment (a more recent prominent example has been Pope Paul's VI anti-birth control encyclica), the hesitation in counteracting the "population bomb" is mainly due to two factors: (1) as human beings we tend to underestimate the dangers of exponential growth. Obviously, we have no reliable imagination of the consequences of an exponential growth rate until we see, literally, the result; (2) reduction of the death rates through effective public health measures without a concurrent reduction of birth rates is considered the prime cause of the rapid increase in population numbers (21). Because birth rates and family planning are intimately interwoven with the social systems and cultures of individual nations, global population *control* is difficult to achieve, if not impossible. In many nations, in particular in the Third World, effective population control within two generations, say, would require a drastic change of venerable propensity structures and value systems. This kind of change was *not* required when measures to decrease death rates were introduced, since in this case the goal structure was simple and readily approved of by almost everybody. It is much more difficult to tell these people that a steady state economy is inevitable and that the indispenable prerequisite for any kind of steady state economy is a zero or slightly negative growth rate of the human populations. How to convince these people that a large family must be considered a sin; that fertility has become a countervalue; that the third child of a family and any further child must be taxed? The present situation is grotesque. As HARDIN (22) noted, "Some spokesmen for poor countries now assert that the poor have the right to produce as many babies as they want, while the rich have the responsibility to keep all these babies alive. No system can be stable when rights and responsibilities are thus separated." How can we, hic et nunc, persuade the governments of the Third World to accept responsibility for the actions and fate of their people and for the world? Let us assume we could really achieve this dramatic change in the propensity structure and value system, world-wide and within the next few generations; let us assume we could in fact stop and even reverse population growth within the next 25 years. Could we then hope that man as a species and human culture as an institution may survive for a prolonged period? Perhaps.

Human culture is based on low entropy structures. Since the supply of fossil neg-entropy is limited (with regard to petroleum *very* limited), we must intensify our efforts to reorient technology and economy toward the only reliable energy source, sunlight. Otherwise, the life of man will be sharply limited by his "terrestrial dowry" of useful low entropy materials. The remaining terrestrial stocks of these materials (including fissile material for atomic energy) must be used predominantly to change technology and economy in a way that a steady state economy, based on sunlight, can be established (I ignore fusion as an energy source, since its technological feasibility is totally uncertain). In principle, the chances for this heroic act are not quite zero: (1) Sunlight is a tremendous energy source. The earth's outstanding recoverable reserves of fossil fuel are estimated to be the equivalent of about two weeks' sunlight (23). (2) Mankind *could* in principle gradually lower population density to a level that could be fed by agriculture at a somewhat *reduced* yield (the reduction of yield is inevitable since the input of inorganic fertilizer must be reduced and part of the land must be set aside to produce hydrocarbons as a replacement for petroleum). (3) New agricultural and forestry technologies could be developed or optimized to grow plants (high-yielding trees, legumes, and tropical grasses such as sugarcane and corn) as raw material for the production of fuels and chemicals (28). The nitrogen contents of these plants could be low. In breeding these new varieties, the goal must be to maximize yield in relation to total biomass. The conventional goal—to maximize yield with regard to usefulness as food material for man to livestock—can be ignored. Even at present where the production of fuel oil, gasoline, and synthetic polymers is based mainly on fossil fuel, the use of agricultural and logging residues as high value raw material is seriously considered (28). There is a large mass of these wastes, and their advantageous properties as fuel are well known. The sulfur content is negligible in comparison with that of coal or crude oil. Likewise, the ash content is reasonably low, and the recovered ash may be used as fertilizer. However, in the foreseeable future there is no realistic chance to base energy, hydrogen, and hydrocarbon production on renewable biomass. In view of the limited and diffuse availability of renewable resources, they can only be regarded as auxiliary fuels and chemical feedstocks rather than as direct substitutes for fossil fuels (28). To replace fossil fuels by renewable resources would require a strong research effort and considerable investments. In every regard, the transition of our present-day culture toward a steady state economy based on solar energy would be a tremendous goal. The realization of this goal would require not only the full power of science and technology and an excessive investment of fossil fuel and fissible material to build up the low entropy structures of the new technologies, but also a strong and lasting change in our propensity structure: soberness, intellectual discipline, renunciation on many levels, the readiness to work hard and consistently, the readiness to abandon forever the conventional illusions about the human right to exploit and degrade the earth, the readiness to relinquish the colorful illusions of the countercultures and citizen activist organizations about idyllic solutions for our present cultural calamity. In any case, a steady state economy based on sunlight as the only energy source would be a labor-intensive economy, which makes full use of human resources, both intellectual *and physical*. SCHUMACHER'S romantic dream of "skilful, productive work of human hands, in touch with real materials of one kind or another" (24) will become harsh reality. We will indeed have to replace many machines by human hands. A strong decrease in our standard of living (as measured by our present-day standards) will be inevitable. This decrease of standards would probably also include medicine and general welfare with a concomitant decrease of life span.

A *positive* factor in the calculation would be our genetic drive to grow in space and time, which originated under the selective conditions of the Paleolithicum. This innate drive could probably be overcome by continuously diverting the intellectual, spiritual, and physical power of mankind to the task of *optimizing* our anticipated economic steady state under conditions of scarcity of energy. One could call this undertaking the "*qualitative* growth" of our steady state culture. At least the emotional reward for an achievement under steady state conditions could be as high or even higher than under conditions of quantitative growth and expansion. In any case, the transition of our present culture to an equilibrium economy must be considered the greatest challenge mankind has encountered so far. And any challenge of this caliber will lead to an utmost activation of man's intellectual and emotional repertory if the goal, the anticipated model to meet the challenge, is attractive and convincing.

I believe that the question of "what will people live and strive for if the situation is stationary? What will be their ideals, will they be bored, will they start fighting each other?" (25) overlooks a major factor, namely, that a steady state culture based on sunlight as the major energy source could not be maintained without a continuous high input of human effort. Thus, the feeling of accomplishment will remain; the satisfaction of having achieved something that is worth the input may be even higher than in our present culture, whose affluence is based on the reckless and almost unconscious waste of fossil neg-entropy.

If we cannot persuade the young generation to give up the monopoly of the present over future generations, it will probably become too late for a controlled change. Once our easily accessible energy bonanza is exhausted, we will no longer have the neg-entropy and thus the technological power to perform in practice the transition to a steady state economy based on sunlight. The fatal collapse of human culture would then become inevitable. We know, of course, that man has to face his ultimate destruction some day. But the point of extinction could be postponed. I agree, of course, that the obstacles seem almost insuperable.

As NICHOLAS GEORGESCU-ROEGEN, a fabulous scholar and a brave economist in favor of a steady state economy said: "Will mankind listen to any program that implies a constriction of its addition to exosomatic comfort? Perhaps the destiny of man is to have a short, but fiery, exciting and extravagant life rather than a long, uneventful and vegetative existence . . . Evolution, even exosomatic evolution, is not reversible—man would rather die in the penthouse than live in the cave" (26). I may add that 95% of present mankind could not return to the caves, even if they wished to, because there are no caves to house mankind. Indeed, cultural evolution has long reached the point of no return.

We have no choice; we must develop a science-based, labor-intensive, highly efficient, disciplined steady state economy or perish. For the sake of our grandchildren I suggest the abandonment, willingly and almost immediately, of the venerable supreme value of quantitative growth in favor of the following commitment: behave in such a way toward your neighbor and toward the environment that even your grandchildren will have a chance to live a decent, humane life on this planet. This commitment, this new supreme value, requires discipline. Discipline in this context is "a system of training, a set of accepted customs or rules that make it possible for a person to reach a state of orderliness through self-control" (27). Self-control is part of maturity and a means to freedom. It leads to freedom, since real freedom implies an insight in necessity and a respect for the necessary limits without compulsion.

Despite the gloomy prospect I have outlined, many scientists have remained optimists in the heart of their heart. Why? The progress of science, the functioning of the ethics of science, the discovery of the natural laws, and the enjoyment of the intrinsic beauty of nature create a positive, optimistic attitude toward human life, toward people, toward the future, and even toward human nature. Certainly, a *pessimistic* view of human nature tends to be self-fulfilling. You can hardly fail if you assume that human nature is rogish in principle, but it is not very pleasant to live with this assumption. This is 'man' as existentialists conceived 'man' to be. "He is angry, and he feels guilt about feeling angry. But he cannot feel joy. Indeed, he cannot even laugh" (1). Existentialism did not pay any serious attention to the rise of science, and scientists, by and large, ignored this pessimistic and joyless wing of contemporary philosophy. The scientific enterprise needs the principle "hope"; science is an *optimistic* adventure of the human mind.

As a scientist deeply concerned about the future of man, one would like to see a renaissance of at least some of those spiritual values without which human life cannot persist in dignity and joy. I am oldfashioned enough to call these indispensable values cardinal virtues: knowledge, integrity, kindness, and diligence; and I call their countervalues cardinal vices: (deliberate) ignorance, corruption, selfishness, and indolence; and I want to define human dignity explicitly: to act as a human being means to act by consideration of values and by conscious decision. The loss of the right to make a decision implies the loss of human dignity. This is the main reason why the modern totalitarian regimes appear so inhumane. I would suggest that consideration and development of values and the learning of the art to make decisions is a prime purpose of colleges and universities. Therefore, I have emphasized the responsibility of the scientist and the intrinsic value system of science in this lecture series.

I want to close this lecture with the words of an ethicist, BARROWS DUNHAM, whose writings have helped me to see essential things more clearly. He said:

The things we do as human beings are of three kinds: we know, we judge, and we enjoy. When we know, we are aware of what is in fact the case. When we judge morally, we are aware what choice is free of bias and considers all people as worth doing things for. When we enjoy with discernment, we are aware of something that engages sense and thought and feeling, that stretches these toward their limits, yet keeps them in harmony with one another and with the world (1).

15th Lecture

Epilogue
Epistemology and Evolution[9]

Einstein once stated that for him the most unintelligible thing about the world is that it is intelligible. Why can we use the axioms and theorems of Euclidian geometry to reason about the physical world? Why is it legitimate to apply to a wheel (a physical object) the mathematical formula derived for a circle

$$c = 2\pi r$$

As you all know, the use of diagrams is not essential to geometry. Geometrical reasoning per se is purely abstract. If diagrams are introduced, it is only as an aid to our reason. In any case, a circle and a wheel are totally different things, but nevertheless the wheel obeys the formula obtained for the circle.

I have written here another formula, derived by Gauss, for a purely mathematical relationship between two variables

$$y = \frac{1}{\sigma\sqrt{2\pi}} \cdot e^{-(x-\bar{x})^2/2\sigma^2}.$$

Why is it possible to describe and treat the frequency distribution in biological populations with the help of this relationship, e.g., the frequency distribution of intelligence test scores in a human population? In brief: why is mathematics, a purely deductive system of axioms and theorems, applicable to nature? Galileo stated in 1623 that "nature is written in mathematical language" (and this phrase has been repeated and followed by scientists ever since) but he could not give any explanation why this is so.

We may extend our question to the whole of logic (I consider mathematics to be part of logic). Why can we rely on syllogistic reasoning?

---

[9]This lecture was delivered at the Department of Philosophy, University of Massachusetts, to a general audience.

It is probably that the *principle* of the syllogism was formulated not before but long after the usefulness and validity of syllogistic reasoning was discovered by man. Logic is the *theory* of deductive argument, not its source. As you possibly remember, WITTGENSTEIN in the *tractatus* defined logic as the study of everything that can be known in advance of experience, everything that is a priori.

Why does the real world obey logic? WITTGENSTEIN (in his early phase) was very concerned about this question. As he puts it, our justification for holding that the world could not conceivably disobey the laws of logic is simply that we could not say of an unlogical world how it would look (1). This is obviously not a good argument. Rather, it is a sign of perplexity and ignorance.

Remember AYER in *Language, Truth and Logic* (2). His second class of propositions were the formal propositions of mathematics and logic, and they were held to be tautologies. AYER thought of them (as did Wittgenstein) "as being merely rearrangements of symbols which did not make any statement about the world." As AYER points out:

> the empirist does encounter difficulty . . . in connection with the truths of formal logic and mathematics. For whereas a scientific generalization is readily admitted to be fallible, the truths of mathematics and logic appear to be *necessary* and *certain* to everyone. But if empiricism is correct, no proposition which has a factual content can be necessary or certain. Accordingly, the empirist must deal with the truths of logic and mathematics in one of the two following ways: he must say either that they are not necessary truths, in which case he must account for the universal conviction that they are; or he must say that they have no factual content, and then he must explain how a proposition which is empty of all factual content can be true, useful, and surprising. If neither of these courses proves satisfactory, we shall be obliged to give way to rationalism. We shall be obliged to admit that there is some truth about the world which we can know independently of experience; that there are some properties which we can ascribe to all objects, even though we cannot conceivably observe that all objects have them. And we shall have to accept it as a mysterious inexplicable *fact* that our thought has this power to reveal to us authoritatively the nature of objects which we have never observed.

I will try to convince you that the fact Ayer is referring to is neither mysterious nor inexplicable. Ayer (and nearly all philosophers so far) did not take into account that experience has been accumulated and preserved as genetic information during genetic evolution. From the point of view of the individual, this inherited foreknowledge about the structure of the world has the character of synthetic judgements a priori; from the point of view of evolution, however, the same statements must be regarded as synthetic judgements a posteriori, based on experience.

Let me briefly state the same problem we have just described on the level of logic and mathematics on the level of sensory impressions and constructs. How could we talk about physical objects if we were only connected to the external world by sensory impressions? How could we build up a picture, a model, of an external world in terms of stable patterns if the information stems exclusively from the unstable and confusing stream of sensory impressions. Our picture of the world is an imaginative projection, the result of an active and creative process; it is not only due to the constancy of sensory routes.

Any physical object is a construct, an intellectual invention, that is useful in organizing the sense impressions. A tree is a construct. In MARGENAU'S (3) words: "The tree is real because it is the rational terminus of certain rules of correspondence having their origin in sense impressions and because it satisfies the demands of consistency which common sense imposes." Do we learn the constructs? Some classes of constructs, yes! We learn what a gene is, an electron, a molecule. But do we learn what a tree is, a man, a mother? I feel that we must have some genetically inherited foreknowledge about the structure of the world with regard to the invention of valid constructs; otherwise we would not be able to organize intellectually the world around us to the extent and with the rapidity a sane human mind actually does.

Inductive inference is the core of scientific discovery (see Fig. 1). Is the creation of a theory a random phenomenon, is the choice of a particular theory a random act? I think not. Rather, it seems to the observer of scientific progress that some kind of foreknowledge about the structure of the world will enable us to advance reasonable theories, bypassing the multitide of possible theories. Inductive inference is amazingly well aimed.

In summing up, I want to remind you of Kant's famous dictum that, although there can be no doubt that all our knowledge begins with experience, it does not follow that it all arises out of experience. Since Kant means the knowledge of the individual, he is correct. Kant had no access to the idea that a process we call "genetic evolution" has necessarily been related to the accumulation of experience over time. Kant could not know that the seemingly inexplicable a priori knowledge of the individual is actually a posteriori knowledge about the world, laid down in the peculiar nucleotide sequence of the DNA in our genes.

Evolution is a universal phenomenon. The universe shows characteristics of evolution, galaxies evolve as do stars and their planets, including the earth. On the level of living systems evolution—the so-called organismic evolution—has the characteristics of genetic evolution, determined—in the case of sexually reproducing organisms—by mutation and recombination—which lead to genetic variability—reproductive fitness, selection, and concomitant adaptation. The factors and laws of organismic evolution differ from the factors and laws of physical evolution. Evolution of a star, e.g., is determined exclusively by physical laws; the theory of organismic evolution requires biologic laws in addition to physical laws, in particular the concept of acquisition and conservation of genetic information. Organismic evolution is considered a fact, in the sense that it is a phenomenon that is no longer doubted by any competent man. The explanatory *theory* of organismic evolution is predominantly based on the statistical laws of population genetics. These laws permit probabilistic statements on random mutation, recombination, and adaptive selection toward fitness. The laws of population genetics are by no means historical laws, since their validity does not depend on their application to the phenomenon of organismic evolution. The laws of population genetics have no history.

The explanation of evolution by the intellectual means of Neo-Darwinism offers no principal difficulties, although the detailed explanation of the evolutionary process, e.g., of the rapid extinction of whole classes of organisms at certain points of evolution, is impossible as long as we do not know the boundary conditions that prevailed by the time of the catastrophic event. You remember that scientific explanation requires not only a knowledge about the pertinent laws, but also an equally reliable knowledge about the pertinent boundary conditions (see Fig. 3).

I will make only a few remarks about the mechanism of evolution; just a few statements that seem to be indispensable for our theme proper. The gene is the unit of inheritance, the individual is the unit of selection, and the population (as a rule, the species) is the unit of genetic evolution. Mutations, i.e., changes of genes, are not particularly rare events; however, it seems that most mutations do not have a significant influence on the phenotype of the individual. Since selection works through the phenotype, it affects only those genes that become expressed in the phenotype. The existence of selectively neutral mutations shows that the principle of natural selection does not apply at the level of genes (DNA) or proteins in the way it does at the phenotypic level.

The more recent concept of "inclusive fitness" deserves particular emphasis (4). It states in principle that an individual's genetic fitness is to be measured not only by the survival and reproduction of himself and his offspring, but by the enhancement of the fitness of other relatives who share his genes. This allows the evolution of cooperative acts that may be detrimental to the individual's own survival or reproduction and are therefore by definition altruistic. The concept of "inclusive fitness" in its quantitative ramifications extends the evolutionary mechanism without any break or discontinuity to the evolution of sociality, including man's unique position among the social species. Just a hint: "Inclusive fitness" is not restricted to a group of individuals who are related by kinship. A "friend," e.g., is a person whose traits and thus whose genes we like. We treat a "friend" as if he were a person with whom we share many common genes. Another example: an old man, who is no longer able to father children is still of importance to the theory of evolution since he may contribute to the "inclusive fitness" of the population. This explains, if the argument is developed in detail, the high prestige that old, experienced, and bright people have enjoyed in human societies.

Macroevolution, a term to designate relatively rapid and dramatic evolutionary events, can be explained by the same mechanism as microevolution, which is directly accessible to the experiments of the population geneticist, if we assume that mutational changes of the central propensity or aim structure of an organism will take the lead. That is to say, only those mutations of the executive organs will be preserved that fit into the general tendencies *previously* established by the changes of the central propensity structure. This idea was originally suggested by Sir KARL POPPER in his famous Herbert Spencer Lecture of 1961 about "Evolution and the Tree of Knowledge" (5). To quote some of POPPER'S lucid and well-aimed examples: "To take an example, this theory would suggest that the specialized beak and tongue of the woodpecker developed, by selection, *after* it began to change its tastes and its feeding habits, rather than the other way round. Indeed, we may say that had the woodpecker developed its beak and tongue before changing its taste and skill, the change would have been lethal: it would not have known what to do with its new organs. Or take a classical Lamarckian example—the giraffe: its propensities or feeding habits must have changed *before* its neck, according to my theory; otherwise a longer neck would not have been of any survival value" (5). In an Addendum to the Spencer Lecture written in 1971, POPPER elaborated the idea further that, our effort to explain macroevolution, we must start with behavioral or ethological mutants, i.e., "organisms whose differences from their parents consist primarily in their deviating *behavior*" (5). A novel behavior, due to a mutational change in the central propensity structure, may greatly increase the selective value of an anatomical structure or a physiologic system already existing in the particular organism. In the famous example of the eye, POPPER points out that (5) "the *novel behavior* which makes use of light-sensitive spots (already existent) may greatly increase their selective value, which previously perhaps was negligible. In this way, *interest* in seeing may be successfully fixed genetically and may become the leading element in the orthogenetic evolution of the eye; even the smallest improvements in its anatomy may be selectively valuable if the aim-structure and the skill-structure of the organism make sufficient use of it."

The idea advanced by Popper has been considered also by *professional* theoreticians of evolution, e.g., by ERNST MAYR (6). The experts confirm that indeed mutational changes of *behavior* may cause new strong selection pressures leading to rapid and dramatic morphological and physiological changes, simulating orthogenetic evolution and permitting the conquest of new territories. In conclusion, large-scale progression in evolution must probably be attributed predominantly to mutational changes in the central propensity structure of organisms.

We now turn to the evolution of man. Evolution of man is a fact.

This statement means that there is unanimous consensus among competent scientists that the gene pool and the gene frequencies of recent human populations originated during a historical process governed in essence by the laws of genetic evolution, including "natural selection." The expression natural selection implies in the theory of evolution that on the average those individuals who are genetically determined to be the best adapted to a given environment will have the highest probability of transmitting their genes to the next generation. This is the meaning of the statement that the fittest show the highest survival rate of progeny. I have already mentioned that "inclusive fitness" extends this concept to societies. Natural selection is a kind of negative feedback. It is, in negative terms, the extinction of error and of poorly adapted gene combinations.

Only recently, during the past few thousand years, has large-scale teleological action, i.e., goal-aimed, deliberate and conscious action leading to the development of culture, interfered seriously with natural genetic evolution in man as well as in the animal and plant kingdoms. In particular the development of culture—including the creation of beliefs and values—has led to a dramatic change on the level of selection. In historical times, natural selection has been replaced more or less by "artificial," man-made selection, as documented by the introduction of agriculture, medicine, and welfare (caring in particular for the aged and infirm members of a society) and by the practice of *active* birth control in some way or other. This statement does not imply that genetic evolution, i.e., changes in gene frequencies, no longer occurs in recent human populations. However, it implies that those factors and laws that have governed the ascent of man in the context of *organismic* evolution have been replaced to some extent by cultural practices. This fact has terminated the phenomenon of *natural* organismic evolution on this planet. On the other hand, however, the deterioration of the original gene pool of the human populations by artificial selection has not yet advanced very far. In essence, our anatomy, physiology, and behavior are still largely determined by those genes that have been selected in course of the natural evolution of man, including those genes that determine our central propensity structure. I do not want to follow this aspect any further, since we need for our present theme ("epistemology and evolution") only the following statements:

1. Natural genetic evolution has led to modern man—to the populations of *Homo sapiens*—and to his major attributes: cognitive properties, *verbal* language as an effective means of communication, description and argument, a highly sophisticated central propensity structure, teleological action. These attributes are bound to the most complex system in nature, the human brain.

In the ontogeny of man the brain is the most elaborate example of developmental homeostasis (7). This means that the build-up of the central nervous system is controlled by intrinsic genetic factors without any *spe-*

cific influence of extrinsic factors. The brain is completed shortly after birth. Every man, except monozygotic twins, has his specific brain as indicated, e.g., by the specificity of the electroencephalogram. Only in genetically identical, *mono*zygotic twins, are the electroencephalograms identical.

2. The gene pool and the gene frequencies of recent human populations are not too different from the gene pool and gene frequencies of our ancestors who lived, say, 6000 years ago. Despite teleological action and concomitant cultural evolution, the genes and gene frequencies of present man may still be considered by and large as the result of an extremely competitive natural genetic evolution toward adaptive fitness in the real world.

We now return to the original question, is the human mind a tabula rasa in the sense of the strict empirist. Must we adhere to the venerable doctrine that knowledge, and especially scientific knowledge, always starts from observation? Or, are there inborn structures, innate ideas, already present at birth? If the latter alternative must be preferred, how can the existence of inborn knowledge be explained?

Aristotle assumed that the axioms of logic would be inborn; Kant thought that a great deal of what we take to be features of objective reality—e.g., space and time and causality are actually manufactured in some sense by the human mind. CHOMSKY holds the thesis that there is an inborn universal grammar (8). We are left with the question whether or not there is any convincing explicit *explanation* for the existence of a kind of *instinctive* foreknowledge about the real world in the untouched human mind.

For a biologist "instinct" is strictly defined. Instincts are *genetically* determined intrinsic, species-specific adaptive patterns of reaction or patterns of behavior that can be released by specific key stimuli from the environment. In man, instinctive patterns of behavior are more abundant than is generally assumed. This observation leads us to the question, are there in fact any innate cognitive structures comparable in some way to inherited instincts. Science can give a straightforward answer to this question: some genetic inheritance of cognitive *properties* is indeed an established fact. The analysis from the point of view of the developmental geneticist is difficult for three reasons.

Firstly, complex and integrative properties such as language or intelligence or temper will be determined simultaneously by many genes. The analysis of multifactorial inheritance of quantitative traits requires a more sophisticated approach than studies in Mendelian genetics, but in principle we know how to handle the problem.

Secondly, as Darwin already remarked, it is manifest that man is subject to much variability. Actually the variability of every trait or character in human populations is considerably higher than in any other natural population. It is a serious problem to disentangle the contribution of genetic variation and environmentally caused variation to the total variation of a trait in a given population. But again, methods have been elaborated to approach this problem, even for human populations. While the whole field of research is still in flux, we know already that under normal environmental conditions—which can be defined—a considerable part, probably a large part, of the *variation* of cognitive properties in human populations is due to genetic variation (9).

Thirdly, in those cases where learning plays a decisive role, the actual performance of a man will be determined by his genes and by the environment in a multistep sequence. Let us consider language as an example. The learning of a language is part of our adaptive behavioral learning. The basic ability to learn and to speak a language depends on a set of basic genes. A lack or mutation in this set of genes leads to genetic diseases, e.g., to a deaf-mute. All men who are healthy with regard to this set of basic genes can learn and speak any natural language. There is good evidence that human speech exhibits features that are common to all natural languages (8).

The type of natural language that a human being learns will be determined completely by the environment, as a rule by the mother: mother tongue.

However, the *quantitative* ability, the capacity, to make use of a mother language and to pick up foreign languages is again determined, at least in part, by genetic factors. Language fluency and language weakness and all the grades in between depend to some extent on the *particular* set of genes we have inherited.

To sum up: a language has to be learned, but the basic potential and the "capacity" are innate. This is probably true for all a priori knowledge. It requires adequate stimulation and the sensory impressions from the normal, "natural" environment of the human being to express and to optimize the innate potential. By "normal" we mean the environment to which we became genetically adjusted in the course of evolution, including the mixed genetic-cultural evolution of our hominid ancestors on their way to the recent *Homo sapiens*.

For a general audience the most convincing and unambiguous examples of the inheritance of cognitive properties are those defect mutants that can be studied through Mendelian genetics. Mental deficiencies due to the lack of a single gene are well-documented in developmental genetics and medicine. We *know* that the difference between Bertrand Russell and an idiot can be due to a single gene.

Why do I mention explicitly those things that you can read in every textbook on developmental genetics? Remember the basic question behind my talk: How does it come that our cognitive structures and the structures of the real world coincide to some extent a priori?

I will now answer this question: the cognitive apparatus of man is the result of genetic evolution. Our subjective cognitive structures coincide to some extent with the objective structures of the real world, because they have developed during genetic evolution as an *adaptation* to the real world. An at least partial coincidence of cognitive structures and real structures has been essential for the survival of man as soon as teleological action, i.e., conscious, goal-aimed, willed, and purposive action came into play. Thinking before acting requires right thinking! The major argument in favor of this answer is that every trait, every character, which depends on genes, is necessarily subject to genetic evolution. If the cognitive properties of man are really inherited, at least in part, there must have been an evolution of cognitive traits. And this evolution must have become fast and seemingly orthogenetic as soon as cognitive power and concomitant teleological action became a selective advantage over purely instinctive behavior. Teleological action that implies not only the conscious anticipation of goals and of means to reach a goal but also of consequences and undesirable side effects (see Fig. 11) must have been accompanied by strong changes in the central propensity structure of the very early man. Man began to think along certain lines before he went into action, at least sometimes. In any case the genetic ascent of man must have been marked by the adaptation of man's cognitive structure to the structure of nature. Otherwise man would not have survived genetic evolution once teleological action was invented. As soon as man started to think in the framework of teleological action, he had the alternatives, think predominantly right—and succeed, or, think predominantly wrong—and perish.

Only a very few human populations, all belonging to the same species, have survived genetic evolution. One may argue that the other branches of our genealogical tree have disappeared mainly because they could not overcompensate the risk of wrong thinking by the advantage of right thinking and reasoning in accordance with nature. It has been extremely risky in evolution to leave the safe but narrow shelter of instinct-controlled behavior. Even our generation, the modern *Homo sapiens*, is by no means safe. We always walk on the verge of extinction. The tremendous wealth of genuine reliable knowledge elaborated by modern science may become overruled any time not only by the decay of values but also by the inflation of wrong thinking, characteristic of our age. I grew up with existentialism, one of the many examples of revolts against reason—and against joy. The alternative for the recent *Homo sapiens* sounds brutal: think and act right or perish. Right means, in accordance with the laws of nature *and* in accordance with the nature of man. For example: if human aggression is something that cannot be avoided—this is the opinion of leading ethologists—it must be taken into account and redirected by conscious teleological action in ways that will prove relatively harmless to the species and the planet.

Adaptive fitness of cognitive traits can easily be documented. Light and sound perception and the temporal resolving power of the human mind are thoroughly investigated examples in this regard. Kant's categories and *Anschauungsformen* (space and time) have been beautifully analyzed by LORENZ under this point of view (10). This kind of analysis shows that the adaptation of our cognitive structures to the structure of the real world is limited. The adaptation covers only a certain, relatively narrow range. The resolving power of our visual sense is about a tenth of a mm in space and about a sixteenth of a second in time. The reason for this severe limitation of our sensory system and cognitive structures is simply that genetic evolution of man depended on the structure and on the resolving power of sense organs that were developed much earlier in animal evolution. All sensory systems that connect our central nervous system and thus our "mind" to the real world had already been developed in the animal kingdom long before the ascent of man to an extent that they could hardly be improved any further. This is the reason why genetic evolution in man has been predominantly an evolution of the brain. However, adaptive genetic evolution of the brain and concomitant cognitive structures has been limited by the efficiency of our sensory systems or, in other words, by the availability of signals, "experience," from the real world.

Science originated from the crafts and from primitive natural philosophy. These disciplines remained within the realm of middle dimensions to which we became directly adapted during evolution. When some scientific discipline invaded the macro- and the microworlds, one could not expect that our terms, our imagination, and the cognitive structures of our mind would work out there as a matter of course. Fortunately, logic and mathematics seem to work everywhere in nature. As HEISENBERG writes: "All mathematical laws would hold also on the distant stars" (15). However, the epistemological problems in quantum physics (11), the horrible consequences of the second law of thermodynamics if applied to the universe, the difficulties with the notion of time and space in the theory of relativity, are particularly intriguing examples of the *limitations* of our cognitive structures. As HEISENBERG (15) noticed, "either we go to the distant stars, or to very small atomic particles. In these new fields, our language ceases to act as a reasonable tool. We will have to rely on mathematics as the only language that remains. I really feel it is better not to say that the elementary particles are small bits of matter; it is better to say that they are just representations of symmetry. The farther we go down to smaller particles, the more we get into a mathematical world, rather than into a mechanical world." In conclusion, our genetically inherited a priori knowledge about space and time covers only the dimensions to which we became adapted during genetic evolution. It is Euclidian geometry to which we are adapted. Non-Euclidian geometry is a system of axioms and theorems that is definitely beyond the realm of our imaginative faculty. However, Riemannian geometry, a kind of non-Euclidean geometry, is applicable to physical space in the context of the general theory of relativity, i.e., when we consider unusually great velocities or distances. Riemannian geometry was worked out as a pure product of the human mind at the end of the nineteenth century. And then, in general relativity, the curvature tensor in Riemannian geometry and the energy momentum tensor in physics joined.

I have been trying to deal with a venerable problem of epistemology, the validity of synthetic judgements a priori, from the point of view of a scientist. I have been trying to answer an epistemological question by means of a scientific theory, the theory of evolution. Evolutionary epistemology [to use the term of CAMPBELL (12) and VOLLMER (13)] is not only compatible with biological facts, including behavior biology, but also with the more recent developments in the fields of perceptive and cognitive psychology. Beyond this, evolutionary epistemology is inevitable! If the theory of evolution is justified (nobody will doubt this) and if we possess inborn and inherited cognitive structures (no expert will doubt this, I guess), then our cognitive structures must have been subject to genetic evolution and adaptation. However, the adaptation is limited; not only because of the limitations and genetic deficiences of our sensory system, but also because of the mutation pressure in those genes that contribute to the build-up and to the maintenance of the most complicated system in nature, the human brain (7). The selection pressure that aims at optimization and uniformity is always counteracted by the mutation pressure that leads to diversity and concomitant loss of optimization. The particularly wide variation in mental performance and the high incidence of disturbances, including mental diseases, indicate how sensitively this structure responds even to slight changes of the genotype. As I indicated previously, *the whole distance* between an intellectual giant such as Einstein or Russell and a hopeless idiot can be accounted for by the mutation of a single gene.

At the end of my talk I want to return to Kant and to Ayer. AYER writes in *Language, Truth and Logic* (2):

> . . . the admission that there were some facts about the world which could be known independently of experience would be incompatible with our fundamental contention that a sentence says nothing unless it is empirically verifiable. . . . the fundamental tenet of rationalism is that thought is an independent source of knowledge, and is moreover a more trustworthy source of knowledge than experience; indeed some rationalists have gone so far as to say that thought is the only source of knowledge. And the ground for this view is simply that the only necessary truths about the world which are known to us are known through thought and not through experience.

From the point of view of evolutionary epistemology, the venerable confrontation of empiricism and rationalism is a fictive problem. In reality it does not exist, since synthetic judgements a priori are also based on experience. The rationalism vs. empiricism debate is a striking example of a philosophical discussion that scientific advance has rendered pointless.

From the point of view of the individual man, a synthetic judgement about the world that we can know independently of experience is a synthetic judgement a priori in a strict sense. However, from the point of view of evolution, the same judgement is a synthetic judgement a posteriori; it is based on experience, namely on the experience of our ancestors, which is preserved and stored in the genetic information, in the peculiar nucleotide sequence of the genetic DNA we have inherited from our parents. Kant's dictum that, although there can be no doubt that all our knowledge begins with experience, it does not follow that it all arises out of experience, can no more be maintained. The fact is that we combine in our individual life two kinds of experiences: the genetically inherited experience of our ancestors and the experience we have made in our personal life, including the experience transmitted to us by *cultural* tradition and social imitation.

Let me close my talk with a personal remark: in my opinion further progress in science and in epistemology will depend on the strong and steady, i.e., not only occasional, interaction of both fields. Epistemology must respect and consider seriously the actual, genuine knowledge elaborated by scientific disciplines. Science has proved that it is capable of producing genuine knowledge, although the theory of how knowledge can actually be obtained, epistemology, may not yet be satisfactory. The scientific disciplines, in particular those fields that have advanced far beyond the realm of common sense, such as classical quantum physics, elementary particle physics, or molecular biology, must consider epistemology as part of their endeavor. A considerable number of scientists have realized that epistemology is part of their trade. Professional philosophers, at least in my home country, are more reluctant, following Wittgenstein at least in this regard, who firmly believed that the dominance of scientific thought since the Renaissance has been a disaster. However, some recent statements look promising. To quote AYER once more from a recent conversation with MAGEE: "If you want a philosopher to be constructive in the sense of not leaving things where they are, then I think you have got to marry him to a scientist. I don't see how one is to effect changes except by changing one's conceptual outlook, and I don't think one can do this except in conjunction with scientific theories. . . . I think that there has again to be a reapproachment of philosophy and science. They diverged in the nineteenth century, partly in consequence of the romantic movement and partly because science got too difficult, and now there are some signs that they are coming together again" (14). I hope AYER will be right in the end.

References

To the Reader

1. HULL, D. L.: Philosophy of reductionism in biology. Nature (London) **257**, 429 (1975)
2. KRÜGER, L. (ed.): Erkenntnisprobleme der Naturwissenschaften. Köln: Kiepenheuer und Witsch, 1970
3. MICHAELIS, A. R., HARVEY, H. (eds.): Scientists in Search of Their Conscience. Proc. Symp. Impact Sci. on Society. Berlin-Heidelberg-New York: Springer, 1973
4. PORTER, G.: The relevance of Science. In: Novas Tendencias em Fotobiologia. Caldas, L. R. (ed.). Suplemento dos Anais da Academia Brasileira de Ciências, Rio de Janeiro, 1973, Vol. XXXXV

lst Lecture: Prologue

1. MOHR, H.: Lectures on Photomorphogenesis. Berlin-Heidelberg-New York: Springer, 1972
2. WALKER, M.: The Nature of Scientific Thought. Englewood Cliffs: Prentice Hall, 1963
3. MEDWEDJEW, S. A.: The Rise and Fall of T. D. Lysenko. New York: Columbia Univ. Press, 1969
4. MOHR, H.: Der prinzipielle Konflikt zwischen Biologie und Marxismus. In: Marxismus-ernst genommen. Szczesny, G. (ed.). Reinbek: Rowohlt, 1975
5. MOHR, H.: Über die Bedeutung der Naturwissenschaften für die Kultur unserer Zeit. Nova Acta Leopoldina, Vol. 37/2. Halle: Deutsche Akademie der Naturforscher Leopoldina, 1973
6. BÜCHEL, W.: Gesellschaftliche Bedingungen der Naturwissenschaft. Munich: Beck, 1975
7. See Plato's pertinent comments and predictions in the Dialogue "Kritias"
8. BRONOWSKI, J.: In: The New York Times, October 18, 1971

9. SZENT-GYORGYI, A.: In: The New York Times, October 23, 1971
10. ROSZAK, T.: The Making of a Counter Culture. London: Faber, 1970
11. HABERMAS, J.: Technik und Wissenschaft als "Ideologie". Frankfurt: Suhrkamp, 1968
12. MASON, H. L.: Reflections on the politicised university I: The academic crises in the Federal Republic of Germany. AAUP Bulletin **60**, 299, 1974
13. JEVONS, F. R.: Science Observed. London: Allen and Unwin, 1973
14. DULLAART, J.: Proposal of general, ethical statement for natural scientists. Acta Biotheoretica **19**, 212, (1970)
15. MOHR, H.: Freiburger Universitätsreden, Neue Folge, **46**, 1969
16. MOHR, H.: Wissenschaft und Wertsystem. Langenbecks Arch. Chir. **334** (Kongressbericht 1973), 18 ff. (1973)
17. BEVERIDGE, W. I. B.: The Art of Scientific Investigation. New York: Vintage Books, Random House, 1950
18. JENSEN, A. R.: Harvard Educational Review **39**, 1–123 (1969)
19. HIRSCH, J.: Jensenism: The Bankruptcy of "Science" Without Scholarship. Winter, 1975, Vol. **25**, No. 1
20. McCORMMACH, R.: On academic scientists in Wilhelmian Germany. Daedalus **103**, No. 3, 157, (1974)

2nd Lecture

1. JEVONS, F. R.: Science Observed. London: Allen and Unwin, 1973
2. CHARGAFF, E.: Building the tower of Babble. Nature (Lond.) **248**, 776 (1974)
3. WEISSKOPF, V. F.: In: Scientists in Search of Their Conscience. Michaelis, A. R., Harvey, H. (eds.). Berlin-Heidelberg-New York: Springer, 1973, p. 194
4. MOHR, H.: Erbgut and Umwelt. In: Die Neue Elite. Oberdörfer, D., Jäger, W. (eds.). Freiburg: Rombach, 1975
5. MOHR, H.: Der prinzipielle Konflikt zwischen Biologie und Marxismus. In: Marxismus-ernst genommen. Szczesny, G. (ed.). Reinbek: Rowohlt, 1975
6. HARRIS, T. A.: I'M OK—YOU'RE OK. New York: Harper and Row, 1969
7. HAGSTROM, W. O.: The Scientific Community. New York: Basic Books, 1965
8. LEOPOLD, A. C.: Games scientists play. BioScience **23**, 590 (1973)
9. KETELAAR, J. A. A.: In: Scientists in Search of Their Conscience. Michaelis, A. R., Harvey, H. (eds.). Berlin-Heidelberg-New York: Springer, 1973, p. 96
10. McCORMMACH, R.: On academic scientists in Wilhelmian Germany. Daedalus **103**, No. 3, 157 (1974)

3rd Lecture

1. HOLTON, G.: On being caught between Dionysians and Apollonians. Daedalus **103**, No. 3, 65 (1974)
1a. ECCLES, J. C.: The discipline of science with special reference to the neurosciences. Daedalus **102**, No. 2, 85 (1973)

2. ZIMAN, J. M.: Public Knowledge. Cambridge: Cambridge Univ. Press, 1968
3. CAMPBELL, N.: What is Science? New York: Dover Publ., 1953
4. WOODGER, J. H.: Biology and Language. Cambridge: Cambridge Univ. Press, 1952
5. ROBINSON, J. T.: The Nature of Science and Science Teaching. Belmont: Wadworth, 1968
6. STEEN, L. A.: Foundations of mathematics: Unsolvable problems. Science **189**, 209 (1975)
7. WHORF, B. L.: Sprache, Denken, Wirklichkeit. Reinbek: Rowohlt, 1963
8. CHOMSKY, N.: Introduction to Adam Schaff: Language and Cognition. New York: McGraw-Hill, 1973
9. SCHAFF, A.: Language and Cognition. New York: McGraw-Hill, 1973
10. CHOMSKY, N.: Biological Foundations of Language. New York: Wiley, 1967
11. SAVIGNY, E. v.: Grundkurs im wissenschaftlichen Definieren. Munich: DTV, 1970
12. MARGENAU, H.: The Nature of Physical Reality, a Philosophy of Modern Physics. New York: McGraw-Hill, 1950
13. DAVIES, J. T.: The Scientific Approach. London: Academic Press, 1973
14. RUSE, M.: The Philosophy of Biology. London: Hutchinson Univ. Library, 1973
15. HULL, D.: Philosophy of Biological Science. Englewood Cliffs: Prentice-Hall, 1974
16. SINNOT, E. W., DUNN, L. C., DOBZHANSKY, T.: Principles of Genetics. New York: McGraw-Hill, 1950
17. MOHR, H.: Lectures on Photomorphogenesis. Berlin-Heidelberg-New York: Springer, 1972

4th Lecture

1. ZIMAN, J. M.: Public Knowledge. Cambridge: Cambridge Univ. Press, 1968
2. The integrity of science: a report by the AAAS Committee on Science in the promotion of human welfare. In: Am. Scientist **53**, 174 (1965)
3. MACDONALD, J. R.: Are the data worth owning? Science **176**, 1377 (1972)
4. COMPTON, W. D.: Letter to Science **190**, 834 (1975)
5. POPPER, K. R.: Conjectures and Refutations. London: Routledge and Kegan Paul, 1963
6. DAVIES, J. T.: The Scientific Approach. London: Academic Press, 1973
7. POPPER, K. R.: The Logic of Scientific Discovery. London: Hutchinson, 1959
8. SIMON, M. A.: The Matter of Life. New Haven: Yale Univ. Press, 1971
9. KUHN, T. S.: The Structure of Scientific Revolutions. Chicago: Univ. Chicago Press, 1970
10. HESSE, M.: The Structure of Scientific Inference. London: Macmillan, 1974
11. CAWS, P.: The structure of discovery. Science **166**, 1375 (1969)

12. MOHR, H.: Wissenschaft und menschliche Existenz. Freiburg: Rombach, 1967
13. adopted from EWING, A. C.: The Fundamental Problems of Philosophy. London: Routledge and Kegan Paul, 1951

5th Lecture

1. HULL, D.: Philosophy of Biological Science. Englewood Cliffs: Prentice-Hall, 1974
2. WALKER, M.: The Nature of Scientific Thought. Englewood Cliffs: Prentice-Hall, 1963
3. POPPER, K. R.: Logik der Forschung. Tübingen: Siebeck, 1969
4. MOHR, H.: Biologie als quantitative Wissenschaft. Naturwiss. Rdsch., **7**, (1970)
5. DAVIES, J. T.: The Scientific Approach. London: Academic Press, 1973
6. RUSE, M.: The Philosophy of Biology. London: Hutchinson, 1973
7. HEMPEL, C. G., OPPENHEIM, P.: Studies in the logic of explanation. Phil. Sci. **15**, 135 (1948)
   HEMPEL, C. G.: Philosophy of Natural Science. Englewood Cliffs: Prentice-Hall, 1966
8. A statement by ERNST MAYR, according to HULL (1), p. 86
9. WILLIAMS, M. B.: Reducing the consequences of evolution: a mathematical model. J. Theoret. Biol. **28**, 343 (1971)
   WILLIAMS, M. B.: The logical status of natural selection and other evolutionary controversies. In: Philosophy Sci. Symp. Dordrecht: Reidel, 1973
10. quoted from SIMON, M. A.: The Matter of Life. New Haven: Yale Univ. Press, 1971, p. 84
11. MAALØE, O.: In: Michaelis, A. R., Harvey, H. (eds.): Scientists in Search of Their Conscience. Berlin-Heidelberg-New York: Springer, 1973, p. 134
12. HEISENBERG, W.: Tradition in science. In: The Nature of Scientific Discovery. Gingerich, O. (ed.). Washington, D.C.: Smithsonian Inst. Press, 1975

6th Lecture

1. STEGMÜLLER, W.: Das Problem der Kausalität. In: Erkenntnisprobleme der Naturwissenschaften (L. Krüger, Hg.). Köln: Kiepenheuer und Witsch, 1970
2. MOHR, H., SITTE, P.: Molekulare Grundlagen der Entwicklung. Munich: BLV, 1971
3. for a more detailed treatment see MOHR, H.: Biologie als quantitative Wissenschaft. Beilage zu: Naturwiss. Rdsch. **7**, 779 (1970)
4. DRUMM, H., MOHR, H.: The dose response curve in phytochrome-mediated anthocyanin synthesis in the mustard seedling. Photochem. Photobiol. 20, 151, 1974
5. SCHOPFER, P.: Experimente zur Pflanzenphysiologie. Freiburg: Rombach, 1970
6. WAGNER, E., BIENGER, I., MOHR, H.: Die Steigerung der durch Phytochrom

bewirkten Anthocyansynthese des Senfkeimlings (Sinapis alba L.) durch Chloramphenicol. Planta **75**, 1 (1967)

7. quoted from LANGE, H., BIENGER, I.: In: MOHR, H.: Biologie als quantitative Wissenschaft. Suppl. Naturwiss. Rdsch. **7**, 779 (1970)
8. KOCHANSKI, Z.: Conditions and limitations of prediction-making in biology. Phil. Sci. **40**, 29 (1973)

7th Lecture

1. HARTMANN, N.: Teleologisches Denken. Berlin: DeGruyter, 1951
   EDWARDS, P., PAP, A. (eds.).: Determinism, freedom and moral responsibility. In: A Modern Introduction to Philosophy. New York: The Free Press, 1972
2. WICKLER, W.: Die Biologie der Zehn Gebote. Munich: Piper, 1971
3. WILSON, E. O.: Sociobiology—The New Synthesis. Cambridge, Mass.: Belknap Press of Harvard Univ. Press, 1975
4. POPPER, K. R.: Evolution and the tree of knowledge. In: Objective Knowledge—an Evolutionary Approach. Oxford: Clarendon Press, 1972
5. SCHLICK, M.: Chapter 7, in: Problems of Ethics. New York: Prentice-Hall, 1939
6. CAMPBELL, C. A.: Is "free will" a pseudo-problem. Mind **60**, 441 (1951)
7. EIBL-EIBESFELDT, I.: Der vorprogrammierte Mensch. Vienna: Molden, 1973
8. TOMPKINS, P., BIRD, C.: The Secret Life of Plants. New York: Harper and Row, 1973
9. SALISBURY, F. B.: The return of superstition. BioScience **24**, 201 (1974)
10. MOHR, H., OELZE-KAROW, H.: Phytochrome action as a threshold phenomenon. In: Light and Plant Development. Smith, H. (ed.). London: Butterworth, 1976
11. HAMILTON, W. D.: The Genetical Evolution of Social Behavior. I. J. Theoret. Biol. **7**, 1 (1964)
    HAMILTON, W. D.: The Genetical Evolution of Social Behavior. II. J. Theoret. Biol. **7**, 17 (1964)
12. GUNNING, B. E. S., ROBARDS, A. W. (eds.): Intercellular Communication in Plants: Studies on Plasmodesmata. Berlin-Heidelberg-New York: Springer, 1976

8th Lecture

1. STENT, G. S.: Molecular biology and metaphysics. Nature (Lond.) **248**, 779 (1974)
2. for an excellent, conventional treatment of the problem I refer to RUSE, M.: The Philosophy of Biology. London: Hutchinson, 1973 and to HULL, D. L.: Philosophy of Biological Science. Englewood Cliffs: Prentice-Hall, 1974. Among the philosophers of science, Ernest Nagel has been a strong believer in the possibility of theory-reduction. See NAGEL, E.: The Structure of Science. London: Routledge and Kegan Paul, 1961

3. KUHN, T. S.: The Structure of Scientific Revolutions. Chicago: Univ. Chicago Press, 1970
FEYERABEND, P. K.: Explanation, reduction, and empiricism. In: Minnesota Studies in the Philosophy of Science. Feigl, H., Maxwell, G. (eds.). Minneapolis: Univ. Minnesota Press, 1962
4. WOODGER, J. H.: Biology and Language. Cambridge: Cambridge Univ. Press, 1952
5. WATSON, J. D., CRICK, F. H.: A structure for deoxyribose nucleic acid. Nature (Lond.) **171**, 737 (1953)
6. JACOB, F., MONOD, J.: Genetic regulatory mechanisms in the synthesis of proteins. J. Mol. Biol. **3**, 318 (1961)
7. POLANYI, M.: Life's irreducible structure. Science **160**, 1308 (1968)
8. HESS, B.: Organization of glycolysis: Oscillatory and stationary control. In: Rate Control of Biological Processes. Symp. Soc. Exp. Biol., Symp. XXVII. London: Cambridge Univ. Press, 1973
9. ESTES, W. K.: Human behavior in mathematical perspective. Am. Scientist, **63**, 649 (1975)
10. quoted from STENT, G. S.: Molecular biology and metaphysics. Nature (Lond.) **248**, 779 (1974)
11. quoted from JEUKEN, M.: The biological and philosophical definitions of life. Acta Biotheoretica **24**, 14 (1975)
12. DOMBROWSKI, H. J.: Lebende Bakterien aus dem Paläozoicum. Biol. Z. **82**, 477 (1963)
13. BOHR, N.: Das Quantenpostulat und die neuere Entwicklung der Atomistik. Naturwissenschaften **16**, 245 (1928)
14. DEWAR, M. J. S.: Quantum organic chemistry. Science **187**, 1037 (1975)
ROBINSON, A. L.: Chemical dynamics: Accurate quantum calculations at last. Science **191**, 275 (1976)
15. HEISENBERG, W.: Tradition in science. In: The Nature of Scientific Discovery. Gingerich, O. (ed.). Washington, D.C.: Smithsonian Inst. Press, 1975
16. LAUFFER, M. A.: Entropy-driven Processes in Biology: Polymerization of Tobacco Mosaic Virus Protein and Similar Reactions. New York: Springer, 1975
17. EIGEN, M.: Self-organization of matter and the evolution of biological macromolecules. Naturwissenschaften **58**, 465 (1971)
18. RYCHLAK, J. F.: A Philosophy of Science for Personality Theory. Boston: Houghton Mifflin, 1968

9th Lecture

1. MOHR, H.: Zur Zielsetzung der Physiologie. Naturwiss. Rdsch. **28**, 154 (1975)
2. PRIGOGINE, I.: Introduction to the Thermodynamics of Irreversible Processes. Springfield: Thomas, 1955
FOX, R. F.: Entropy reduction in open systems. J. Theoret. Biol. **31**, 43 (1971)
3. MOHR, H.: Lectures on Photomorphogenesis. Berlin-Heidelberg-New York: Springer, 1972

4. MITRAKOS, K., SHROPSHIRE, W. (eds.): Phytochrome. New York: Academic Press, 1972
5. MOHR, H.: Untersuchungen zur phytochrominduzierten Photomorphogenese des Senfkeimlings (Sinapis alba L.). Z. Pflanzenphysiol. **54**, 63 (1966)
6. SCHÄFER, E., MOHR, H.: Irradiance dependency of the phytochrome system in cotyledons of mustard (Sinapis alba L.). J. Mathemat. Biol. **1**, 9 (1974)
7. SCHÄFER, E., LASSIG, U., SCHOPFER, P.: Photocontrol of phytochrome destruction in grass seedlings. The influence of wavelength and irradiance. Photochem. Photobiol. **22**, 193 (1975)
8. SCHÄFER, E.: A new approach to explain the "high irradiance responses" of photomorphogenesis on the basis of phytochrome. J. Mathemat. Biol. **2**, 41 (1975)
9. ETZOLD, H.: Der Polarotropismus und Phototropismus der Chloronemen von Dryopteris filix-mas (L.) Schott. Planta **64**, 254 (1965)
10. HAUPT, W.: Localization of phytochrome in the cell. Physiol. Végétale **8**, 551 (1970)
11. OELZE-KAROW, H., MOHR, H.: An attempt to localize the threshold reaction in phytochrome-mediated control of lipoxygenase synthesis in the mustard seedling. Photochem. Photobiol. **23**, 61 (1976)
12. SMART, J. J. C.: Philosophy and Scientific Realism. London: Routledge and Kegan Paul, 1963
13. WOODGER, J. H.: Biology and Language. Cambridge: Cambridge Univ. Press, 1952
14. WILLIAMS, M. B.: Reducing the consequences of evolution: A mathematical model. J. Theoret. Biol. **28**, 343 (1971)
15. RUSE, M.: The Philosophy of Biology. London: Hutchinson, 1973
16. HULL, D. L.: Philosophy of Biological Science. Englewood Cliffs: Prentice-Hall, 1974
17. in: GINGERICH, O. (ed.): The Nature of Scientific Discovery, p. 211, Washington, D.C.: Smithsonian Inst. Press, 1975
18. OSCHE, G.: Die Vergleichende Biologie und die Beherrschung der Manningfaltigkeit. Biologie in unserer Zeit, **5**, 139 (1975)
19. MAYR, E.: Principles of Systematic Zoology. New York: McGraw-Hill, 1969
20. MAYR, E.: Systematics and the Origin of Species. New York: Columbia Univ. Press, 1942
21. quoted from M. RUSE (15), p. 125
22. quoted from M. RUSE (15), p. 125
23. MUNSON, R.: Studies in History and Philosophy of Science **5**, 73 (1974)
24. MOHR, H., SITTE, P.: Molekulare Grundlagen der Entwicklung. Munich: BLV, 1971
25. SOKAL, R. R., SNEATH, P. H. A.: Principles of Numerical Taxonomy. San Francisco: Freeman, 1963
26. quoted from M. RUSE (15), p. 161
27. quoted from M. RUSE (15), p. 148
28. HOLTON, G.: Thematic Origins of Scientific Thought: Kepler to Einstein. Cambridge, Mass.: Harvard Univ. Press, 1973
29. WATSON, J.: The Double Helix. New York: Atheneum, 1968

10th Lecture

1. ZIMAN, J. M.: Public Knowledge. Cambridge: Cambridge Univ. Press, 1968
2. KUHN, T. S.: The Structure of Scientific Revolutions. Chicago: Univ. Chicago Press, 1970
3. SIMON, M. A.: The Matter of Life. New Haven: Yale Univ. Press, 1971
4. JEVONS, F. R.: Science Observed. London: Allen and Unwin, 1973
5. LEOPOLD, A. C.: Games scientists play. BioScience **23**, 590 (1973)
6. CRICK, F.: Central dogma of molecular biology. Nature (Lond.) **227**, 561 (1970)
7. CHARGAFF, E.: Building the tower of Babble. Nature (Lond.) **248**, 776 (1974)
8. RAVETZ, J. R.: In: Problems of Scientific Revolution. Harré, R. (ed.). Oxford: Clarendon Press, 1975
9. quoted from BONDI, H.: In: Problems of Scientific Revolution. Harré, R. (ed.). Oxford: Clarendon Press, 1975
10. HEISENBERG, W.: In: The Nature of Scientific Discovery, p. 235. Gingerich, O. (ed.). Washington, D.C.: Smithsonian Inst. Press, 1975
11. BOWERS, R.: The peer review system on trial. Am. Scientist **63**, 624 (1975)
12. MACLANE, S.: Peer review and the structure of science. Science **190**, 617 (1975) GUSTAFSON, T.: The controversy over peer review. Science **190**, 1060 (1975)
13. SNOW, C. P.: The Two Cultures: and a Second Look. London: Cambridge Univ. Press, 1963
14. WEISSKOPF, V. F.: In: Scientists in Search of Their Conscience. Michaelis, A. R., Harvey, H. (eds.). Berlin-Heidelberg-New York: Springer, 1973, p. 200
15. SCHOLEM, G.: In: Scientists in Search of Their Conscience. Michaelis, A. R., Harvey, H. (eds.). Berlin-Heidelberg-New York: Springer, 1973, p. 156
16. HAGSTROM, W. O.: The Scientific Community. New York: Basic Books, 1965
17. STENT, G.: The Coming of the Golden Age: A View of the End of Progress. New York: National History Press, 1971
18. TOULMIN, S.: In: The Nature of Scientific Discovery. Gingerich, O. (ed.). Washington, D.C.: Smithsonian Inst. Press, 1975, p. 198
19. SAMUEL, D.: In: Scientists in Search of Their Conscience. Michaelis, A. R., Harvey, H. (eds.). Berlin-Heidelberg-New York: Springer, 1973, p. 144
20. SATTLER, R.: Towards a more adequate approach to comparative morphology. Phytomorphol. **16**, 417 (1966)
21. SATTLER, R.: Essentialism in plant morphology. Proc. Intern. Congr. History Sci. Tokyo, 1974, Vol. III, p. 464
22. TROLL, W.: Vergleichende Morphologie der höheren Pflanzen. Berlin: Bornträger, 1937
23. ZIMMERMANN, W.: Die Phylogenie der Pflanzen. Stuttgart: Fischer, 1959
24. SATTLER, R.: A new conception of the shoot of higher plants. J. Theoret. Biol. **47**, 367 (1974)

11th Lecture

1. HAGSTROM, W. O.: The Scientific Community. New York: Basic Books, 1965
2. SAYRE, A.: Rosalind Franklin and DNA. New York: Norton, 1975

3. KLUG, A.: Rosalind Franklin and the double helix. Nature (Lond.) *248*, 787 (1974)
4. PEKERIS, C. L.: In: Scientists in Search of Their Conscience. Michaelis, A. R., Harvey, H. (eds.). Berlin-Heidelberg-New York: Springer, 1973, p. 49
5. POPPER, K. R.: In: Problems of Scientific Revolution. Harré, R. (ed.). Oxford: Clarendon Press, 1975
6. ZIMAN, J. M.: Public Knowledge. Cambridge: Cambridge Univ. Press, 1968
7. HEISENBERG, W.: In: The Nature of Scientific Discovery. Gingerich, O. (ed.). Washington, D.C.: Smithsonian Inst. Press, 1975, p. 231
8. KATZIR-KATCHALSKY, A.: In: Scientists in Search of Their Conscience. Michaelis, A. R., Harvey, H. (eds.). Berlin-Heidelberg-New York: Springer, 1973, p. 160
9. PETER, W. G.: GSA members vote on heredity, race, and I.Q. BioScience **25**, 417 (1975)
10. quoted from (9), p. 418
11. LOEHLIN, J. C., LINDZEY, G., SPUHLER, J. N.: Race Differences in Intelligence. San Francisco: Freeman, 1975
12. quoted from (9), p. 465
13. MEDWEDJEW, S. A.: The Rise and Fall of T.D. Lysenko. New York: Columbia Univ. Press, 1969
    DEWITT, S.: Freedom of inquiry. Science **189**, 953 (1975)
14. ARENTS, J.: Controversial Areas of Research. Science **190**, 324 (1975)
15. SAMUEL, D.: Science and the control of man's mind. In: Scientists in Search of Their Conscience. Michaelis, A. R., Harvey, H. (eds.). Berlin-Heidelberg-New York: Springer, 1973
16. DAVIS, B.: In: The Nature of Scientific Discovery. Gingerich, O. (ed.). Washington, D.C.: Smithsonian Inst. Press, 1975, p. 603
17. BERG, P., BALTIMORE, D., BRENNER, S., ROBLIN, R. O., III, SINGER, M. F.: Asilomar Conference on Recombinant DNA Molecules. Science **188**, 991 (1975)
18. RUSSELL, C.: Biologists draft genetic research guidelines. BioScience **25**, 237 (1975)
19. WADE, N.: Recombinant DNA: NIH group stirs storm by drafting laxer rules. Science **190**, 767 (1975)
20. quoted from (18)
21. quoted from (8), p. 58
22. SHINN, R.: In: The Nature of Scientific Discovery. Gingerich, O., (ed.). Washington, D.C.: Smithsonian Inst. Press, 1975, p. 598

12th Lecture

1. TOULMIN, S.: The end of the Copernican era? In: The Nature of Scientific Discovery. Gingerich, O. (ed.). Washington, D.C.: Smithsonian Inst. Press, 1975
2. MOHR, H.: Science and responsibility. In: Lectures on Photomorphogenesis. Berlin-Heidelberg-New York: Springer, 1972

3. MICHAELIS, A. R., HARVEY, H. (eds.): Scientists in Search of Their Conscience. Berlin-Heidelberg-New York: Springer, 1973, p. 51
4. DAVIS, R.: Technology as a deterrent to dehumanization. Science **185,** 737 (1974)
5. EDSALL, J. T.: Scientific freedom and responsibility. Science **188,** 687 (1975)
6. SAMUEL, D. in (3), p. 144
7. CURLIN, J. W.: Mutatis mutandis: Congress, science and law. Science **190,** 839 (1975)
8. quoted from KANTROWITZ, A.: Controlling technology democratically. Amer. Sci. **63,** 505 (1975)
9. DAVIES, J. T.: The Scientific Approach. London: Academic Press, 1973
10. HEISENBERG, W.: Tradition in science. In: The Nature of Scientific Discovery. Gingerich, O. (ed.). Washington, D.C.: Smithsonian Inst. Press, 1975
11. JEVONS, F. R.: Science Observed. London: Allen and Unwin, 1973
12. BROOKS, H.: Are scientists obsolete? Science **186,** 501 (1974)
13. LANGVISH, J., GIBBONS, M., EVANS, W. G., JEVONS, F. R.: Wealth from Knowledge. London: Macmillan, 1972
14. KORNBERG, A.: National Institutes of Health, Alma Mater. Science **189,** 599 (1975)
15. KNOX, W. T.: Letter to Science **190,** 834 (1975)
16. FLOWERS, B. H.: Science, industry and government. Nature (Lond.) **222,** 421 (1969)
17. EHRLICH, P. R., EHRLICH, A. H., HOLDREN, J. P.: Human Ecology. San Francisco: Freeman, 1973
18. MEADOWS, D. H., MEADOWS, D. L., RANDERS, J., BEHRENS, W. W., III,: The Limits to Growth. New York: Universe Books, 1972
19. R. ARON in (3), p. 122
20. SALOMON, J.-J.: Science and scientists' responsibilities in today's society. In: (3)
21. HOLDEN, C.: Hazel Henderson: Nudging society off its Macho trip. Science **190,** 862 (1975)
22. SCHUMACHER, E. F.: Small is Beautiful. London: Harper and Row, 1973.
EDWARD GOLDSMITH, who is editing the "Ecologist" (a kind of a propaganda paper) advocates the dismantling of industrial economies and the return to a rural society embodying small-scale technology and rustic virtues; see WADE, N.: Edward Goldsmith: Blueprint for a de-industrialized society. Science **191,** 270 (1976)
23. WEISSKOPF, V. F.: in (3), p. 197/198
24. WADE, N.: Third World: Science and technology contribute feebly to development. Science **189,** 770 (1975)

13th Lecture

1. NISBET, R.: Knowledge dethroned. The New York Times Magazine: September 28, 1975, p. 34
2. CASSIDY, H. G.: Knowledge, Experience and Action: An Essay on Education. New York: Teachers College Press, Columbia Univ., 1962

3. CASSIDY, H. G.: Scientific thought: A force in human life. Am. Scientist **58**, 476 (1970)
4. SNOW, C. P.: The Two Cultures: and a Second Look. London: Cambridge Univ. Press, 1963
5. HOLTON, G.: Mainsprings of scientific discovery. In: The Nature of Scientific Discovery. Gingerich, O. (ed.). Washington, D.C.: Smithsonian Inst. Press, 1975
6. ROSZAK, T.: The Making of a Counter-Culture. Garden City, N.Y.: Doubleday, 1969
   ROSZAK, T.: The monster and the titan: science, knowledge and gnosis. Daedalus **103**, No. 3, 17 (1974)
7. SCHUMACHER, E. F.: Small Is Beautiful. London.: Harper and Row, 1973
8. quoted from (5), p. 203
9. PALM, A.: Human Values in Science. BioScience **24**, 657 (1974)
10. BRONK, D.: quoted from ABELSON, N. M., ABELSON, P. H.: Detlev Bronk. Science **190**, 941 (1975)
11. BROOKS, H.: Are scientists obsolete? Science **186**, 501 (1974)
12. HIRSCH, J.: Jensenism: The Bankruptcy of “Science” Without Scholarship. Winter, 1975, Vol. XXV, No. 1
13. COTGROVE, S.: Objections to science. Nature (Lond.) **250**, 764 (1974)
14. PEKERIS, C. L.: The impact of physical sciences on society. In: Scientists in Search of Their Conscience. Michaelis, A. R., Harvey, H. (eds.). Berlin-Heidelberg-New York: Springer, 1973
15. in (14), p. 159
16. MARTINO, J. P.: Science and society in equilibrium. Science **165**, 769 (1969)
17. HOLTON, G.: On being caught between dionysians and apollonians. Daedalus **103**, No. 3, 65 (1974)
18. WEINBERG, A. M.: Science in the public forum: Keeping it honest. Science **191**, 341 (1976)

14th Lecture

1. DUNHAM, B.: Ethics—Dead and Alive. New York: Knopf, 1971
2. E. O. WILSON in People Weekly, November 17, 1975
3. ROKEACH, M.: The Nature of Human Values. New York: Free Press, 1973
4. SCHWARTZ, S.: A survey of guiding principles. Science **186**, 436 (1974)
5. BRONOWSKI, J.: Science and Human Values. New York: Messner, 1956
6. LORENZEN, P.: Normative Logic and Ethics. Mannheim: Bibliographisches Institut, 1969
7. CLARKE, B.: The causes of biological diversity. Sci. Am. **233**, 50 (1975)
8. GINGERICH, O. (ed.): The Nature of Scientific Discovery. Washington, D.C.: Smithsonian Inst. Press, 1975, p. 563
9. MAGEE, B.: Modern British Philosophy. New York: St. Martin’s Press, 1971, p. 165
10. SIMON, M. A.: The Matter of Life. New Haven: Yale Univ. Press, 1971

11. LOEHLIN, J. C., LINZEY, G., SPUHLER, J. N.: Race Differences in Intelligence. San Francisco: Freeman, 1975
12. DAVIS, B. D.: Social determinism and behavioral genetics. Science **189,** 1049 (1975)
13. MOHR, H.: Erbgut und Umwelt. In: Die neue Elite. Oberndörfer, D., Jäger, W. (eds.). Freiburg: Rombach, 1975
14. SCHOECK, H.: Der Neid—Eine Theorie der Gesellschaft. Freiburg: Alber, 1968
15. HACKERMAN, N.: Higher education: Who needs it? Science **190,** 513 (1975)
16. MOHR, H.: Wissenschaft und Wertsystem. Langenbecks Arch. Chir. 334 (Kongressbericht 1973), 18 (1973)
17. MEADOWS, D. H., MEADOWS, D. L., RANDERS, J., BEHRENS, W. W., III: The Limits to Growth. New York: Universe Books, 1972
18. L. B. BROWN in The New York Times, December 7, 1975
19. PIMENTAL, D., DRITSCHILO, W., KRUMMEL, J., KUTZMAN, J.: Energy and land constraints in food protein production. Science **190,** 754 (1975)
20. HARDIN, G. (ed.): Population—Evolution—Birth Control. San Francisco: Freeman, 1964
21. FREEDMAN, R., BERELSON, B.: The human population. Sci. Am. **231,** 31 (1974)
22. HARDIN, G.: Letter to BioScience **25,** 148 (1975)
23. WADE, N.: Nicholas Georgescu-Roegen: Entropy the measure of economic man. Science **190,** 447 (1975)
24. SCHUMACHER, E. F.: Small Is Beautiful. London: Harper and Row, 1973
25. WEISSKOPF, V. F.: In: Scientists in Search of Their Conscience. Michaelis, A. R., Harvey, H. (eds.). Berlin-Heidelberg-New York: Springer, 1973, p. 196
26. quoted from (23)
27. FARNSWORTH, D. L.: Social values in college and university. Daedalus **103,** 297 (1974)
28. SARKANEN, K. V.: Renewable resources for the production of fuels and chemicals. Science **191,** 773 (1976)

15th Lecture: Epilogue

1. WITTGENSTEIN, L.: Tractatus Logico—Philosophicus. London: Paul, Trench Trubner and Co., 1922
2. AYER, A. J.: Language, Truth and Logic. London: Gollancz, 1936
3. MARGENAU, H.: The Nature of Physical Reality, a Philosophy of Modern Physics. New York: McGraw-Hill, 1950
4. WILSON, E. O.: Sociobiology—The New Synthesis. Cambridge, Mass: Belknap Press of Harvard Univ. Press, 1975
   HAMILTON, W. D.: The genetical evolution of social behavior. II. J. Theoret. Biol. **7,** 1, 17 (1964)
   TRIVERS, R. L., HARE, H.: Haplodiploidy and the evolution of the social insects. Science **191,** 249 (1976)
5. POPPER, K. R.: Evolution and the Tree of Knowledge. In: Objective Knowledge—An Evolutionary Approach. Oxford: Clarendon Press, 1972

6. a recent statement by E. MAYR ("Wie weit sind die Grundprobleme der Evolution gelöst?") has been summarized in Naturwiss. Rdsch. **27**, 437 (1974)
7. ECCLES, J. C.: Das Gehirn des Menschen. Munich: Piper, 1975
8. CHOMSKY, N.: Sprache und Geist. Frankfurt: Suhrkamp, 1970
9. for a recent summary see H. MOHR: Erbgut und Umwelt. In: Die neue Elite. Oberndörfer, D., Jäger, W. (eds.). Freiburg: Rombach, 1975
10. LORENZ, K.: Kants Lehre vom Apriorischen im Lichte gegenwärtiger Biologie. Blätter für deutsche Philosophie **15**, 94 (1941)
    LORENZ, K.: Die angeborenen Formen möglicher Erfahrung. Z. Tierpsychologie **5**, 235 (1943)
11. HEISENBERG, W.: Die Entwicklung der Deutung der Quantentheorie. In: Erkenntnisprobleme der Naturwissenschaften. Krüger, L. (ed.). Köln: Kiepenheuer und Witsch, 1970
12. CAMPBELL, D. T.: Evolutionary Epistemology. In: The Philosophy of K. R. Popper. Schilpp, P. A., (ed.). La Salle: Open Court, 1974
13. VOLLMER, G.: Evolutionäre Erkenntnistheorie. Stuttgart: Hirzel, 1975
14. MAGEE, B.: Modern British Philosophy. New York: St. Martin's Press, 1971, p. 64
15. HEISENBERG, W.: In: The Nature of Scientific Discovery. Gingerich, O. (ed.). Washington, D.C.: Smithsonian Inst. Press, 1975, p. 557

Index